CHAKRABORTY

THE HAMLYN
ANIMAL
ENCYCLOPEDIA

First published 1975
Second impression 1976
Third impression 1978
Designed and produced by Artia for
The Hamlyn Publishing Group Limited
London New York Sydney Toronto
Astronaut House, Feltham, Middlesex, England
© Artia 1975
© this edition:
The Hamlyn Publishing Group Limited 1975
Illustrations © Květoslav Hísek, Vlastimil Choc,
Vladimír Javorek and František Severa
ISBN 0 600 33536 4
1/12/01/51-03
Printed in Czechoslovakia by TSNP Martin.

The Hamlyn Animal Encyclopedia

DR JIŘÍ FELIX

Edited by Joyce Pope B. A.

HAMLYN

LONDON — NEW YORK — SYDNEY — TORONTO

CONTENTS

FOREWORD

Man's urge to understand the world about him is a heritage from the days when, as a nomadic hunter, his success and survival depended on his knowledge of the movements of the herds of animals which were his prey. The migrations of whole populations of birds and mammals at different times of the year must have been noted and for example, his awareness of the presence of fishes, such as salmon, which could be easily caught at certain places at particular times of the year might have meant the difference between plenty if his observations and predictions were correct and starvation if they were wrong. He must have noticed many such phenomena, and added them to the vast, complicated picture of the natural environment in which he lived.

As man's fund of knowledge and tradition increased, so did his mastery over the natural world, which his cleverness all too often helped him to destroy. Today in Europe we can see small remnants only of the vast natural panoramas which were the homelands of our ancient ancestors. In Britain, for instance, there is no natural landscape as such, for man has modified by tree felling and drainage even the wildest parts of the country. The large wild mammals — elk, bear, wild boar and wolf have long since been extinct there and many of the others, such as martens, polecats and otters have been driven from their ancient haunts. The same picture may be seen with birds, some of which have disappeared entirely, while many others, once common, are now at best extremely rare visitors.

The situation is not, however, entirely black, for man's knowledge of the animal kingdom, once based largely on simple observations, is now becoming augmented by a deeper understanding of what he sees. True, there are still many mysteries — for example we know a great deal about the migration routes chosen by many species, but our understanding of how they navigate on their long journeys is still rudimentary. Many suggestions have been put forward to account for the ability of small birds to hold a true course as they fly over open oceans, but none seems to fill all of the observed facts. In time the problem will be solved, for many approaches are being made in all parts of the world. Perhaps when the answer is found it will illuminate some facet of our own way of life. An example of this happening is to be seen in the behavioural studies on many animals which have shown that patterns of aggression in other animals and man are to a large extent similar. This simple fact could not be comprehended while it was thought that birds sang for joy in the springtime, rather than to assert their territorial rights, to take a well-known aspect of the problem. We may have much to learn from other animals. The old idea of 'Nature red in tooth and claw' dies hard, but modern research has shown that most animal communities live in peace, under a system where a strong heirarchy ensures order.

5

The beginning of our understanding of the animal kingdom came with the realisation of the physical relationships which exist within it. Adaptations to different ways of life may mask these in some cases, but we now know that bats, for example, which were long thought to be related to birds, because of their power of flight, are in fact mammals, with which they share much more fundamental similarities. We are aware that each species has evolved in its environment according to complex but logical laws, which allow the survival of forms most efficiently adapted to their ways of life. The relationships of the animals are expressed in the formal Latin names. These consist of two parts, which taken together are unique to each creature, but the first or generic name is shared by closely related species. An example may be seen in the Tiger Beetles. Three species, *Cicendela sylvatica, C. campestris* and *C. hybrida* are illustrated in this book. They are closely similar, but each has marked preferences for habitat and food which sets it apart from the others.

Animals may also be classified geographically. Europe falls into the vast area known to zoologists as the Palaearctic Region, which also includes much of temperate Asia, where the fauna is substantially different from that of the Nearctic Region of North America, or the Oriental Region of south-east Asia, for example. Man has transported many animals beyond their natural boundaries so we find the American Grey Squirrel inhabiting Britain, which is in a different faunal region from the one in which it originated and many other examples are noted in this book. Classification may also be according to the way of life of the animal. The fact that they live in distinct communities, each with its particular food supply and habits fitting into the whole pattern has increased our comprehension of the world about us and enables us to predict its actions and in some cases control it more sensibly.

In this book the animals of the European part of the Palaearctic Region are described in an arrangement based on the types of habitat found within the area. Thus animals of woodlands, running water or mountains will be found grouped together. Within each section the entries are arranged in an order with the simplest creatures such as worms and other invertebrates coming first, followed by the vertebrates with the birds and mammals at the end of each section. It would not be possible to include all of the animals of Europe in a single volume. In Britain alone about 20,000 species of insect are known to exist; the number for the whole continent is very much greater. A book containing them all would be unwieldly, but it is hoped that in the selection made a representative example of the fauna of each habitat will have been shown. In many cases, especially with the invertebrates, information is lacking on some aspects of the life history or behaviour of the animal. It is hoped that this book will encourage some people to look more closely at the creatures with which we share the world and perhaps by doing so may fill some of the gaps in our knowledge of our environment and ourselves.

THE
FOREST

INTRODUCTION

Unbroken forests once stretched across Europe, and even today large areas remain an important aspect of the landscape. In the past they teemed with game animals which were hunted as a vital food source, and today despite the extinction of the wild oxen and the reduction of the bison to a few protected survivors, several species of deer and the wild boar are found abundantly in some places. Apart from food, the forests have provided man with timber for building, wood for fuel, bark for tanning and resin for various industrial uses and beyond this fodder and shelter for his domestic animals. More important than all these, however, is the fact that forests actually influence the climate of a region and affect the moisture content of the soil so that conditions for life may be affected for a wide area beyond the trees.

Woodlands may be classified in a number of ways. According to their geographical position they may be classed as mountain or lowland forests. Depending on the trees themselves, they may be referred to as deciduous (broad-leaf), coniferous (cone bearing), or of mixed type. Further subdivision is possible according to the species, so that we may have an oak, beech or hornbeam wood among the broad-leaves, although there are usually subsidiary species such as the lime, birch or maple present to break the pure stand of one kind of tree. Coniferous forests more usually contain only one species over a wide area, but even here, the great stretches of fir, pine, spruce or larch woods may be broken by other species where the aspect is more favourable to them.

Deciduous forest vegetation usually grows on three levels. First, at ground level, there is the herb layer, which includes a number of plants mostly flowering in the Spring, before the light is cut out by the leaves growing above them. Anemones, bluebells, lilies of the valley and Solomon's Seal are found among them. Besides this, the herb layer may include various ferns, mosses and liverworts and many fungi, such as the Fly Agaric and species of Boletus which provide food for the larvae of various beetles and flies and also some snails and mammals.

The second level of vegetation, found only where the forest is not too dense, is called the shrub layer and consists of small trees such as hazel or hawthorn which do not reach more than about 6 m. above the ground. The third level is the canopy which consists of the leafy tops of the tallest trees, generally all growing to an equal height in any one area. Each level has its associated fauna, and few animals are found in more than one of them.

Many of the coniferous forests of Europe have been planted by man. Large areas containing only one species may enable the forester to grow a crop which is easily harvested, but the appearance is monotonous and such monocultures are more prone to attack by pests than are the natural forests. These display the same pattern of layers of life as are found in the deciduous woodlands with heather, bilberries, (or in the more mountainous regions bearberry),

8

ferns, mosses and lichens in the ground layer and including juniper in the shrub layer. Most coniferous trees have strong habitat preferences. Scots pines are often found on lowland sandy areas although they will thrive also on exposed high moorland sites while spruce and firs are more characteristic of mountainous regions.

Woodland clearings are of particular biological interest, for any area open to sunlight is quickly colonised by a range of plants not found locally elsewhere. They include willowherb, foxgloves, Aaron's rod, lupins and later wild raspberries. Insects, reptiles and birds are attracted to the area but within a few years seedlings from nearby trees have grown sufficiently to shade out this luxurient growth and the forest takes over once more. The wild life of woodlands is specialised to the habitat. Ancient areas of forest have a rich fauna often differing considerably according to the type of forest concerned. The following pages describe in greater detail where some of these animal occur and how they live.

MOLLUSCS

TWO-LIPPED DOOR SNAIL —
Lacinaria biplicata

This snail is found in damp open woodlands in Central and Eastern Europe, but is rare in Britain, occurring in a few localities only in southern England and at Cambridge. Until the last century it was found at Chelsea and Hyde Park in London, but seems now to have died out there.

The Two — lipped Door Snail usually keeps to moist and shady places, either in thick vegetation, on tree trunks or even on rocks and boulders. Although quite small (the spindle-shaped pale horn-coloured shell is about 17 mm. long when fully grown) it is easily recognised by the regular strong ridges which adorn it. This species is viviparous producing living young and is said to survive for four or five years.

CRAVEN DOOR SNAIL —
Clausilia dubia

The Craven Door Snail is a fairly uncommon species in Britain occurring mainly in northern England although it is widespread in Scandinavia and Central Europe in both mountain and lowland areas. The shell grows to about 16 mm. high. It is smooth and slightly shiny and sometimes may be streaked with white. Like the previous species it prefers moist places, and may be found on rocks and tree trunks, most usually in woods growing in limestone areas. It lives for several years, hibernating throughout the winter months in the leaf litter of the forest floor.

BLACK SLUG —
Arion ater

This slug, which may measure up to 20 cm. long varies greatly in colour, and red, yellow and white forms have been described. Features by which it may be recognised include its rather stout overall appearance, the transverse lines on the foot-fringe and its habit when alarmed of humping itself up into a hemispherical attitude. It is found throughout Britain and Europe from north Scandinavia to Corsica, eastwards into Poland and has now been introduced into New Zealand. It is not solely a woodland creature but also inhabits fields, hedgerows, gardens and wild places feeding on leaves, fruit, fungi and dead animal and plant material. Black slugs are hermaphrodite and all inividuals are capable of egg laying. Mating occurs in the Autumn and eggs are laid throughout the year, taking up to 7 weeks to hatch.

DUSKY SLUG —
Arion subfuscus

The Dusky Slug may grow to a length of **8 cm.** in its two years of life but is usually rather smaller than this. It is a fairly common woodland and meadow species found almost throughout Britain, and on the Continent it extends from Central Europe to North Cape and also east into Siberia. It has been introduced by man into the United States and New Zealand. A useful identification feature is the yellow colour of the body mucus although the overall colour is dusky brown. It feeds on many kinds of soft vegetation and decaying matter, but especially on fungi.

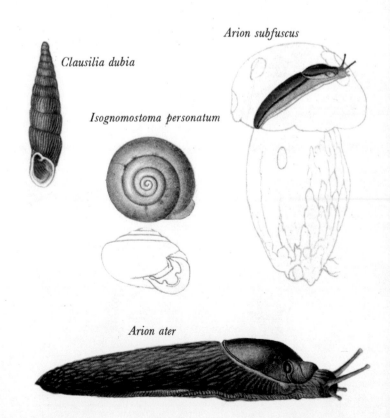

Clausilia dubia

Isognomostoma personatum

Arion subfuscus

Arion ater

Isognomostoma personatum

This snail lives among boulders and under fallen trees in mountain forests in Southern and Central Europe but does not occur in Britain. The shell, which may be up to 6 mm. high and 10 mm. wide is semi-transparent and has three tooth-like projections on the lip. *Isognomostoma personatum* is reddish brown in colour and covered with small fine hairs. Where present it may be extremely abundant and its shell may be found in large numbers among the stones of the hillside.

SPIDERS (and others)

Argiope bruennichi

In this handsome spider the female may reach a size of 15 mm., although her mate, which is an inconspicuous brown colour, is only 4 mm. long. Common in Europe, it has recently been introduced into Britain where several thriving colonies now exist. The large web of Argiope bruennichi is usually found in low vegetation on the edges of woods or in clearings and is decorated by two broad zig-zag bands of silk, which probably help to disguise the presence of the spider. After mating the male is usually eaten by his mate; if he escapes he seeks another female from which encounter he is unlikely

to survive. The female makes a large egg cocoon, which is hung among grasses near the web, the young emerging the following Spring.

Meta segmentata

This spider is common throughout Britain and much of Europe where its characteristically inclined web is pitched in low vegetation in a wide variety of habitats. The species is very variable, both in colour and in structure for there are two broods, one in Spring and the other in Autumn. These two seperate broods differ slightly from each other. *Meta segmentata* is unusual in that the male is as large as the female and that as he waits on the edge of her web his courtship actions are triggered by the arrival of an insect which she kills and which is then the focal point of their mating behaviour.

Xerolycosa nemoralis

The Wolf Spiders, such as *Xerolycosa*, do not build webs, but are active hunters, chasing small ground living prey. A number of similar species exist, most of them specialised in their habitat.
This particular species which lives in forests is known from southern England and across Europe. Females may be easily recognised by the relatively large egg sac which they carry attached to their spinnerets. Like many spiders they tend their young devotedly and care for them until they are independent.

SHEEP TICK —
Ixodes ricinus

Ticks are parasitic animals closely related to mites and like them the adult has eight legs. This is a characteristic which they both share with the spiders. The Sheep Tick is extremely common over much of Europe and the British Isles. It may be found on the edges of woods, in scrubland or on open rough pastures and will parasitise not only sheep, but also a wide range of other animals, including human beings. They lie in wait upon low vegetation and when a suitable host passes by they clamber on to

Argiope bruennichi　　*Glomeris marginatus*

Xerolycosa nemoralis

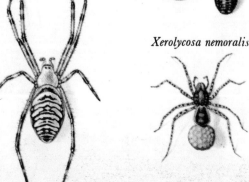

it with remarkable speed. The male is small and can take little blood; the female, which is about 6 mm. long, has a very distendable body and can take a large meal during which her weight may increase many times.

Such a feast takes a considerable period and the tick may spend several days attached to her host. She holds on with her deeply buried mouthparts and also with the adhesive pads and claws on her legs. Any attempt to remove the tick by pulling will normally tear it in two leaving the mouthparts embedded in the host. When the meal is complete, the animal will drop off and she then lays a large number of small red eggs which hatch into six-legged larvae which complete their development in about eight months. This involves two larval changes with blood meals in between them.

Ticks produce a substance called ixodin which acts as a local anaesthetic and prevents the host from noticing any pain or discomfort from the bite. Unless the infestation is really very severe the amount of blood lost is normally negligible, but the Sheep Tick can carry diseases, the most serious of which is encephalitis which is inflammation of the brain.

MILLIPEDES

SNAKE MILLIPEDE —
Iulus sp.

There are many species of Snake Millipede in Britain and Europe but to identify them exactly is the work of a specialist. However, they may be recognised in general by their very long, cylindrical bodies, which are divided into a number of similar segments, each bearing two pairs of legs. They have rather long antennae, which seem to be their principal sense organ, for their eyesight is at best extremely limited.

Snake Milipedes live in the soil under stones and may sometimes be found climbing trees but are more usually ground animals, chiefly active at night. When disturbed they can curl up into a flat coil, making themselves look rather like a watch spring. They are protected against enemies by a series of glands

Lithobius forficatus

Ixodes ricinus (and female)   *Iulus* sp.

on their sides which produce an unpleasant smelling and tasting secretion. The food of all millipedes is soft or rotten vegetation so that in gardens they may be pests feeding on seedlings and the roots of delicate plants.

PILL-BUG —
Glomeris marginatus

The Pill-bug is a short bodied millipede, up to about 14 mm. long but very broad in comparison to its length. At first sight it looks like a Wood Louse and like some of these it can roll itself up into a tight ball when danger threatens. Like all millipedes it has two pairs of legs which are attached to each body segment. It feeds largely on decaying vegetation in woodland soil over most of Britain and Europe, hiding under bark or stones during the time when it is inactive. Pill-bugs have a heavier shell than most of their relatives and so they are less liable to desiccation and may sometimes be found in comparatively dry places.

CENTIPEDES

BROWN CENTIPEDE —
Lithobius forficatus

Centipedes have only one pair of legs to each body segment and are quick moving carnivorous animals using poison to subdue their prey unlike the slow,

15

plant-eating millipedes. Like them, their bodies are not fully waterproofed and so they must live in damp places in the soil and under stones and bark or face the chance of desiccation and death in the dry world of open spaces. They are most usually found in the soil and under stones, always singly, for two would fight.

The Brown Centipede is a chestnut coloured animal 20—30 mm. long and is widespread in Britain and Europe. It has a somewhat flattened body and rather longish legs which stick out at the side. The young have fewer segments than the adults; when first hatched they have seven pairs of legs, but this number increases with each moult over several years of growth until it has the adult number of fifteen pairs.

INSECTS

PINEAPPLE GALL caused by —
Adelges abietis

Pineapple Galls are curious knobbly swellings on the tips of shoots of spruce trees which may be seen throughout Europe, Britain and America. These are caused by attacks of one of the species of Conifer Woolly Apids such as *Adelges abietis*. The life cycles of these tiny insects may be very complex involving reproduction by sexual and asexual means, but the gall is caused by the feeding activities of the first generation of the Spring. This damages the plant tissues in such a way that they swell and protect the insect, although they later open up and release winged aphids which fly to other trees for further feeding.

FOREST BUG —
Pentatoma rufipes

This shield bug is found throughout most of Britain and much of Europe in deciduous forests, orchards and gardens. Oak and Alder are its principal hosts although those living in orchards also often feed on caterpillars, which they suck dry using the strong proboscis. Forest Bugs are present through most of the summer months, breeding chiefly in August and hibernating high in the tree tops as third stage nymphs, finally completing their life cycle during the next year. However, in a good year some may complete their development, overwinter as adults and be present in this form early during the next season.

SNAKEFLY —
Raphidia notata

Snakeflies are not commonly seen, but they are easily recognised by the great elongation of the front part of the thorax, which makes them appear to have a long neck and also by their membranous wings, folded in an inverted 'V' over the abdomen. The female has a long needle-like ovipositor.

This species, which is about 30 mm. long may be seen in British woodlands searching up and down tree trunks for spiders and various other small invertebrates which are its food. The larva, which hides under bark is slim and fast moving and quite as voracious as the adult. Like the adult it has a very long prothorax; a characteristic which enables it to be distinguished from some beetle grubs which it otherwise resembles.

GIANT WOOD WASP —
Urocerus gigas

One of the most fearsome looking of European insects appears in coniferous woods from middle to late Summer. This is the Giant Wood Wasp, one of a group of wood-boring insects which are quite harmless to man although they may be regarded as forest pests, for the larvae burrow into the wood of living trees.

At the end of the abdomen, the females carry a long tube which looks like a sting, but is in fact used solely for egg laying. With this they bore a hole to a depth of several centimetres. into the trunk of pine or spruce trees and deposit a small batch of eggs, laying in all over 300 in different places. When these hatch the grub begins to feed in the wood filling the tunnel behind it with a mixture of wood dust and excrement. It lives in this safely for two or three years when it pupates and finally bites its way free. Because of the length of larval life these insects may complete their development in timber used in buildings where they make ghostly scratching noises as they near emergence. The adult, which flies by day, has a relatively short life.

ICHNEUMON FLY —
Rhyssa persuasoria

This strange looking insect, which in spite of its name, is a relative of the bees and wasps, may be seen in pine or spruce woodlands during the Summer. The 30 mm. long body of the female has an ovipositor 35 mm. on the end of it making this the largest British Ichneumon. She may be seen tapping the tree trunks with her antennae searching for the tunnels of Giant Wood Wasps. Once the correct spot has been found, the female raises her abdomen and inserts her ovipositor into the wood and lays her eggs on or very near to the body of the woodwasp larva.

When they hatch the grubs immediately feed on the still living Wood Wasp, not killing it until their development is complete and biting an escape hatch for themselves after emerging from pupation. These insects feed only on Wood Wasps and keep their numbers in control. Sometimes when Wood Wasps get carried to places where they have no natural enemies as in New Zealand, *Rhyssa* is imported to reduce them.

CHERRY GALL —
Cynips quercus-folii

Many plants respond to the attacks of insects by producing galls which are swellings of the leaves or stems in which they are protected and fed. The plants usually include several different species of rose and also herbaceous plants. One which is found in Britain and Europe is the Cherry Gall of oak leaves.

This gall is 20—25 mm. in diameter, yellowish or red in colour and has a hard outer skin and firm juicy interior containing a small central chamber which is the home of a legless grub. In Autumn this pupates and a reddish wasp-like insect emerges

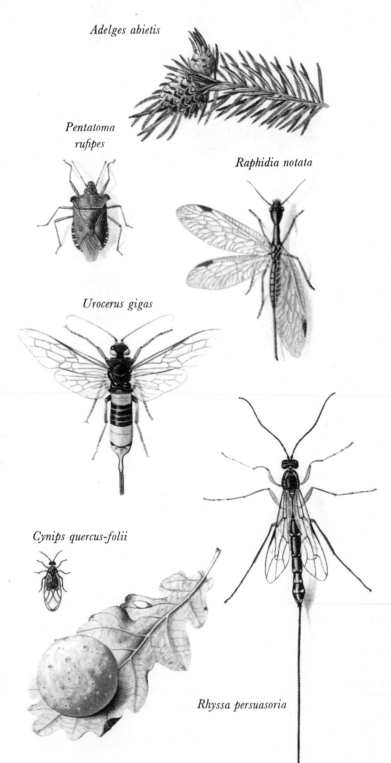

Adelges abietis

Pentatoma rufipes

Raphidia notata

Urocerus gigas

Cynips quercus-folii

Rhyssa persuasoria

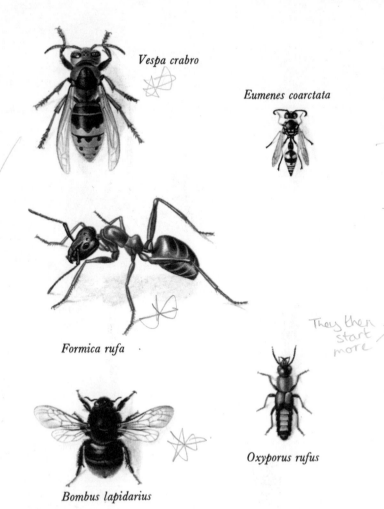

Vespa crabro

Eumenes coarctata

Formica rufa

Oxyporus rufus

Bombus lapidarius

the queen Hornet wakes from her hibernation and searches for a suitable place to make a nest. The most favoured site is in a hollow tree where there will be shelter from rain but other sites including birds' nest boxes have been used.

Wasps' nests are made of a papery material manufactured by the insect from wood scraped with its mandibles from fence posts and other dead, but not rotten wood. This wood is mixed together with saliva and applied to the upper surface of the nest area. From here a stalk is formed and the first cells are hung from this. These open downwards and the eggs are prevented from falling out by being stuck to the side. The grubs when they hatch are fed by the queen until, after pupation, they become worker Hornets. These then take on the task of further nest building and tending the brood, while the queen confines herself to laying more eggs. Wasps in the early stages of development are carnivores and the larvae are fed on the flesh of insects caught by the foraging workers.

During the early part of the Summer when the colony is growing fastest, Hornets and other wasps destroy thousands of insects which would otherwise be harmful to forests, farms and gardens. It is only quite late in the season when there are large numbers of workers and fewer grubs to feed that they become a nuisance, for the adults need sugars to fuel their flight so they attack fruit and any other sweet substances. The majority of Hornets in the nest are sexless females, incapable of mating, but towards the end of the season the queen lays eggs which hatch into fully sexed males and females. After mating, the females which will be the queens of next year find a secure place for hibernation. The males and the rest of the colony, which may have numbered several thousands, all die. Hornets and other wasps possess a sting but they are generally reluctant to use it. Although unpleasant it is rarely seriously harmful to humans.

although it remains in hiding until early Spring. These insects are always female but they are able, nonetheless, without mating, to lay eggs in the leaf-buds of the oak trees and these quickly hatch into tiny insects, which are only 2 mm. long, and of both sexes. After mating the females lay their eggs on the veins of the developing leaves of the oak and it is as a result of these that the Cherry Galls are formed. Very many gall-forming insects have this curious alternation of generations in the course of the year.

HORNET —
Vespa crabro

The largest of British wasps is the Hornet which is found chiefly in southern England, although it is native to much of Europe as well. In springtime

POTTER WASP —
Eumenes coarctata

An insect found widely in European forests, and on heathlands in Britain is the Potter Wasp. This

18

curious looking creature, which is only about 15 mm. long, builds cells of mud and saliva which are well attached to low growing shubby plants. These are the nurseries for its young, but the Potter Wasp does not feed and tend them throughout their larval stage as the grubs of Hornets or other social wasps are cared for. Instead each cell is provisioned by the female with a store of small caterpillars which she immobilises with her sting. She lays an egg in each cell and the grub then finds itself hatching in a larder of moribund food which is sufficient for its growth.

WOOD ANT —
Formica rufa

Hillocks frequently more than a metre in height, made of pine needles, small twigs and leaves may often be found in coniferous woods in Britain and Europe. These are the fortresses of Wood Ants, one of the most abundant and important animals of the forest.

Radiating from the hills are trackways which are the paths taken by the ants in their foraging expeditions when they hunt for insects, worms, snails or any other small animal which they can subdue and carry back to the nest. They do not confine their hunting to the ground but climb and run over all the trees in the vicinity of their home, clearing them of almost all harmful insects. The ants like sweet things as well and their main source of sugar is the honeydew secreted by aphids which are carefully tended and protected and milked by the ants.

The queen ant which is the mother of the entire nest may be up to 11 mm. long, in contrast to the majority of its occupants, the workers, which are up to 9 mm. long; but they may be much less. The workers build the nest and guard it against attack, biting with their strong jaws and squirting formic acid, which may make up a large proportion of their weight, at any intruder. In spite of their ferocity in the face of attack, the workers also tend the eggs, grubs and pupae of the nest, carrying them to the most suitable spots and feeding the larvae with the flesh of animals that they have gathered.

Each year a number of fully sexed males and females are produced and these swarm from all the nests over a large area on a suitable, warm, humid day. This is the mating flight, after which the males die but the fertilised females return to the ground, shed their wings and start their lives as egg producers. They may found new nests but they often join established colonies so that it is not uncommon to find Wood Ant societies with several queens in a single hill.

Camponotus ligniperdis

This large black ant with red-brown legs inhabits decaying tree trunks, especially spruces, in Europe, but it does not occur in Britain. The females, which may be seen during the swarming period, are extremely glossy with dark wings.

The colonies which these ants found include a soldier caste, easily recognised by its large head and powerful jaws, which will defend itself, if picked up, by biting ferociously. Occasionally these ants may form a nest in living wood when the maze of galleries which they chew out will cause considerable damage to the tree.

BUMBLE BEE —
Bombus lapidarius

There are many kinds of Bumble Bee, which may be difficult to identify positively. This particular one is a common species found over most of Britain and Europe.

The queen, which is about 25 mm. long, emerges in late Spring from her winter hibernation and starts to seek out a suitable place for her nest. An old mouse nest which is approached by a long tunnel is the most favoured place. In the soft bedding material used by the mouse she makes her first wax cell and provisions it with pollen and on this lays her first eggs.

These eggs hatch into helpless grubs and are tended by the queen until after pupation. They are then ready to emerge as workers which take over the task of enlarging the nest and rearing the young while the queen restricts herself almost entirely to egg

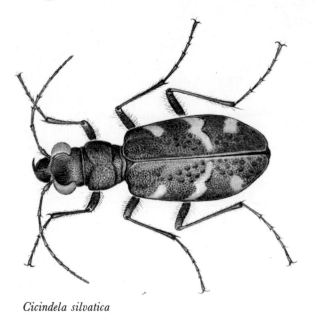

Cicindela silvatica

laying. The nest is enlarged over and around the original cell and it ultimately contains about 300 workers.

Towards the end of the Summer fully sexed males and females are produced and after this no more working brood is reared. The males and females mate, probably with individuals from other nearby nests and after this all but the fertilised females die. These find a secure place for hibernation, for they will be the queens and founders of next year's colonies.

WOODLAND TIGER BEETLE —
Cicindela silvatica

Tiger Beetles are all voracious predators of other invertebrates. This species, which is fairly widespread in Europe, is on the edge of its range in Britain where it occurs locally only in southern England. It usually lives in wooded areas on loose and sandy soil and may be seen on warm days in the midsummer months running rapidly over the ground and taking off abruptly for short fast powerful flights.

The large eyes of the adult enable it to see its prey which it seizes in its secateur-like jaws. The larva is equally carnivorous but rather less active. It digs a tube in the ground which may be up to one metre deep. By means of a curious hump on its back it is able to jam itself near the opening of the hole and from there to grab any small insect which may try to approach. If any danger threatens the Tiger Beetle can drop to safety at the bottom of the hole. It has a very long larval life, hibernating during the Winter. It spends only about a month in the pupal stage, finally emerging as an adult beetle in the late Summer, but not breeding until the subsequent year.

GOLDEN GROUND BEETLE —
Carabus auronitens

This beautiful beetle is not native to Britain, but is found underneath the bark of trees which grow in mountainous regions of Europe. It is most active at night, hunting many kinds of invertebrates. Like the Tiger Beetle the Golden Ground Beetle survives for more than one season. During the wintertime it hibernates in old tree stumps.

CATERPILLAR HUNTER —
Calosoma sycophanta

Of the many beetles which are useful to man, few are more handsome than the Caterpillar Hunter. Although only about 30 mm. long it feeds voraciously on caterpillars and during the two or three years of its life may destroy several hundreds of them. The larva is highly active and climbs over the branches and twigs of trees looking for its prey, but pupates in a safe hole in the ground, which is where the adult also passes the winter months. In years when caterpillars such as those of the geometric moths are specially abundant, the caterpillar hunter will also thrive and by its activity keep the pests in check. Its value is recognised in some places and it is a protected species. Although it is common throughout Europe in all types of forest, only rare immigrants reach Britain and it is not established there. Several very closely related species leading a similar sort of life to this are found in America.

Calosoma inquisitor

This beetle, which is a near relative of the Caterpillar Hunter, occupies a similar niche but is active rather earlier in the year. It is found particularly in oak woods, where its main food is the Looper Caterpillars which are abundant in the spring months. If, as sometimes happens, there is an unusually large number of these defoliators, then the result is that *Calosoma inquisitor* will also be very abundant and may be seen scuttling along forest paths or over tree trunks. This beetle is occasionally found in Britain but is a southern species and does not occur as far north as Scotland.

ROVE BEETLE —
Oxyporus rufus

The family of Rove Beetles is a very large one and the European fauna contains over 3,000 species of them (see plates). They are easily recognised as the hard wing cases do not entirely cover the abdomen, which is soft and unprotected. The larva of this species feeds on forest-living fungi and therefore accounts for the worm-eaten appearance of some toadstools.

SEXTON BEETLE —
Necrophorus vespillo

One of the most valuable of the functions of many insects is as scavengers removing the bodies of animals and the remains of plants. The Sexton Beetles are particularly important in this respect, and some such as *Necrophorus vespillo* are widespread in Britain and Europe. They are found under small dead animals, usually those in a fairly advanced stage of decomposition and to which they may be attracted from a considerable distance.

Although the beetles themselves are only about 20 mm. long they burrow underneath the corpse of the animal and may scratch away the soil to a depth of as much as 30 cm., finally burying it. They lay their eggs on the corpse and the larvae feed on the rotting flesh and pupate in the ground nearby.

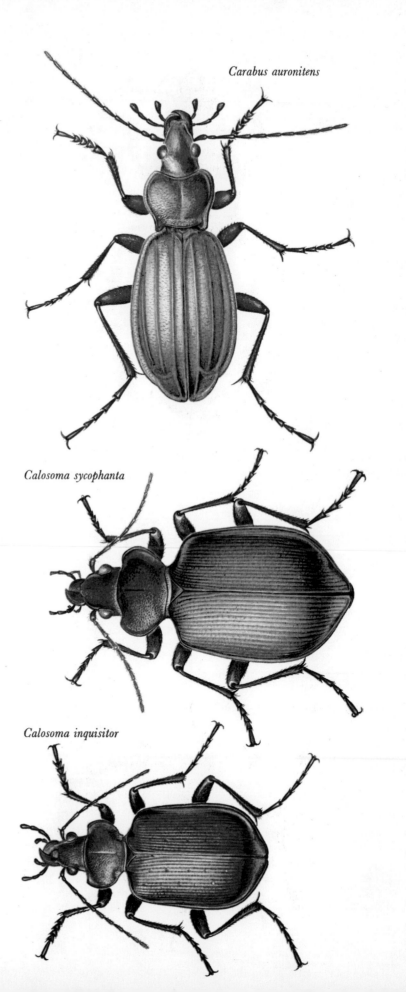

Carabus auronitens

Calosoma sycophanta

Calosoma inquisitor

Necrophorus vespillo

Oiceoptoma thoracica

Necrophorus germanicus

This species which is about 25 mm. long is one of the larger burying beetles. It is widespread in Europe, but does not breed in Britain, although it has been recorded in some southern counties. Its size and its red-tipped, spade-shaped antennae make it easily recognised among all the other European burying beetles. A very similar species which is common in Britain is called *Necrophorus humator*. These beetles are found on rather larger carrion than are the previous species.

BLACK CARRION BEETLE —
Silpha atrata

This beetle, which is fairly common throughout Britain and Europe is usually found associated with small dead animals in rather open places, such as footpaths or the edges of woods. As well as vertebrates it also feeds on dead snails and if food is very short, it may become an active hunter of snails or worms. The Black Carrion Beetle protects itself from predators by giving off an extremely foul smelling secretion.

FOUR-SPOTTED CARRION BEETLE —
Xylodrepa quadripunctata

This beetle, which is about 12—14 mm. long is widespread throughout Europe but is uncommon in Britain occurring only in the midlands and southern England. It is technically a member of the Carrion Beetle group but unlike most of its relatives it is an energetic hunter feeding mainly on caterpillars, largely those of the Black Arches Moth and the Loopers which are forest pests.

The Four-Spotted Carrion Beetle hunts its prey in the trees, as do the curious flat-shaped larvae which survive for the whole Summer, although the adults die in about June after the first flush of caterpillars is over. The larvae pupate in the soil in late Summer but the adult, which is formed by the Autumn, does not emerge as an active predator until next Spring.

Oiceoptoma thoracica

This strikingly coloured beetle, which is about 12 mm. long is to be found throughout Britain and Northern Europe. It is a carrion feeder associated with the carcasses of small animals in deciduous and mixed forests.

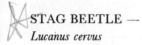

STAG BEETLE —
Lucanus cervus

Stag Beetles, the males of which measure up to 75 mm. in length, are the largest British beetles. They are normally found only in the south of the country, associated, as in the rest of Europe, with oak trees. Stag Beetles fly at night, their wings making a loud whirring noise and their only food during this phase of their life is just the sap of the oak trees.

The female, which is a good deal smaller than her mate, lays her eggs in rotting oak stumps. The larva, which is white and helpless, feeds and develops here for from three to five years and during that time may grow to a length of 100 mm. after which it fashions a small chamber in the soil for pupation. By this time it is possible to determine the sex of the creature, for the males have already developed their large mandibles or 'antlers'.

The specimen figured is typical of the European beetles; the British form has simpler antlers, but they seem to be equally effective when two males are fighting for a mate. In spite of their fierce appearance they are quite harmless to man.

LESSER STAG BEETLE —
Dorcus parallelopipedus

This beetle, which varies in size between 20 and 32 mm. long, can be seen through the summer months in deciduous woodlands throughout Europe, including the Mediterranean, although in Britain it is restricted to the southern counties. It flies late in the day and after mating, the female lays her eggs in the rotting stumps of various broad-leafed trees, where the larvae develop.

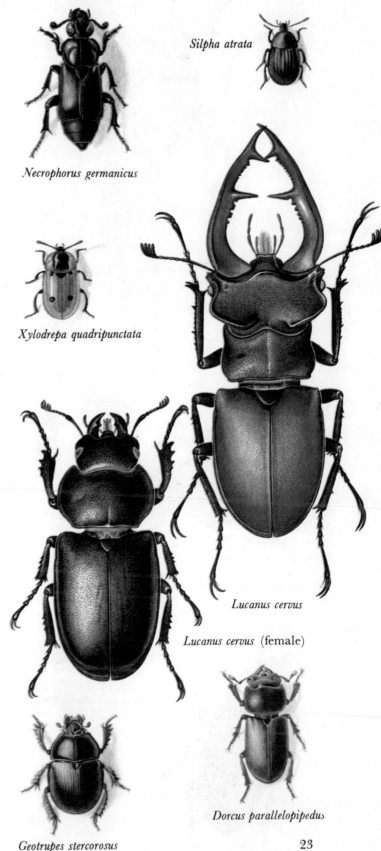

Necrophorus germanicus

Silpha atrata

Xylodrepa quadripunctata

Lucanus cervus

Lucanus cervus (female)

Dorcus parallelopipedus

Geotrupes stercorosus

23

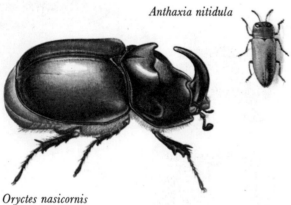

Polyphylla fullo

Anthaxia nitidula

Oryctes nasicornis

Osmoderma eremita

Chalcophora mariana

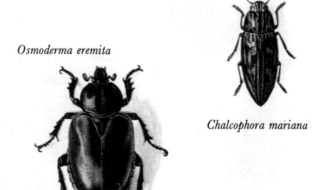

DUMBLE DOR —
Geotrupes stercorosus

Dumble Dors are common forest beetles whose strange buzzing flight, which gives them their English name, may be heard throughout the summer months.

The adults feed on dung and various other rotting substances, but the larvae are solely dung eaters. The female, when about to lay eggs, searches for the droppings of a forest herbivore such as a deer or rabbit and excavates a tube which may be up to 60 cm. deep, beneath them. She then digs out numerous side tunnels and provisions the whole complex with dung which she drags down with her broad front legs, backing into the tunnels and finally filling them almost completely. When the eggs hatch the grubs need not stir from their safe hiding place, making a pupal case from the uneaten remains of their larval meal. Although they emerge before the end of the Summer they hibernate and are the reproductive generation of the next year.

FULLER —
Polyphylla fullo

This handsome beetle, measuring up to 36 mm. in length, is to be found in mainland Europe from Spain to Sweden, chiefly in pine forests near to large rivers and along the sea coasts. It has been recorded from Britain, but always in coastal areas in circumstances which make it likely that it has been accidentally imported, rather than having flown across the channel for itself.

The adults, which are active during the summer months, feed on leaves and as well as defoliating the pines, may also turn their attentions to grape vines. The larvae, which take three or more years to complete their development, live in the soil and feed on the roots of grasses.

RHINOCEROS BEETLE —
Oryctes nasicornis

Rhinoceros Beetles are not recorded from Britain, although they are widespread in Europe and are found across Asia into India. The species gets its name from the large spike on the head of the male which is absent in the female. More variable in size than many insects male Rhinoceros Beetles may measure as little as 20 mm. or as much as 40 mm. in length.

The larvae feed on decaying wood and other rotting vegetation. Adults may sometimes be found in greenhouses having been carried in with compost at an earlier stage of their lives, but since they are nocturnal they may not always be noticed at once.

Osmoderma eremita

This beetle which measures between 24 and 30 mm. long is known in Germany as the Hermit. It is usually found in broad-leafed forests where there are plenty of decaying tree stumps of oak, beach or lime. The eggs are laid in these rotting tree stumps and the larva lives for several years, growing to a length of 100 mm.

The adults are very active at night during the early Summer, and if handled will defend themselves with an unpleasant smell. Although widespread in Central, Southern and Eastern Europe this beetle does not occur in Britain.

Chalcophora mariana

Many European beetles are beautifully coloured, but few are as gaudy as *Chalcophora mariana*. It is easier to see than many other species, for not only is it large, measuring from 25—30 mm. in length, but it is active during the daytime. It is often to be seen basking in the sun on the trunks of pine trees. The larvae feed on dead wood. The species is fairly widespread in Europe, however it does not occur in Britain.

Anthaxia nitidula

This beautiful small beetle, which only measures about 6 mm. in length is to be seen in open glades on the edge of woodlands where the flowers of shrubs and herbaceous plants offer it food. It is widespread in Europe but is a rare species in Britain, occurring in the

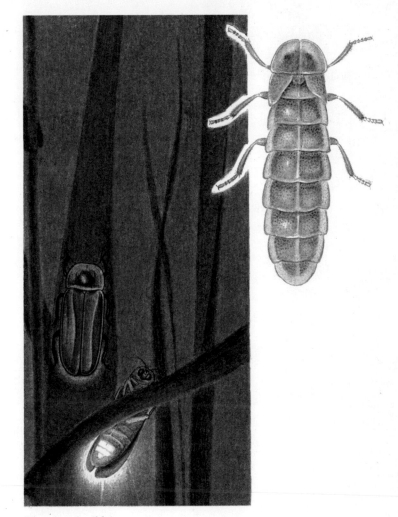

Phausis splendidula

New Forest and other southern woodlands. Other members of the same family have become stored product pests, especially in grain and dried fruits.

Phausis splendidula

The names 'fire fly' and 'glow worm' have been given to many kinds of beetle. This species which is found abundantly in some parts of Europe, does not occur in Britain and differs from the British luminescent beetle in that the male, female and larvae all have light producing organs.

Like the British glow worms, however, the female is wingless and the male homes in to her greenish luminescence. The light is the result of chemical reactions within the body, produces no heat, and can be easily controlled by the insect which can dim its lamp at will. The larvae of *Phausis* are carnivorous, feeding on snails and other invertebrates. The adults, which are rather short lived, probably usually feed on nectar.

EYED LADYBIRD —
Anatis ocellata

This beetle which measures up to 8 mm. in length is the largest British species of ladybird. The dull orange elytra carry the 'eyes' –- black spots surrounded with a white line. The Eyed Ladybird is an inhabitant of coniferous woodlands where it preys on aphids during its larval and adult life and is an important control of these pests.

Thanasimus formicarius

This beetle is up to 10 mm. long, but is rarely seen in Britain although it has been reported as occurring widely there. It is known throughout the continent of Europe and Northern Asia wherever there is a suitable woodland habitat. Its presence depends on bark beetles on which it feeds.

The female lays her eggs on tree trunks and when they hatch the larvae immediately crawl into the tunnel of such insects as *Ips typographica* (see page 30), feeding at first on the grass which they find there but later becoming active predators of the bark beetle, both larvae and adult. After pupation the adult *Thanasimus formicarius* continues its life as a hunter of these forest pests.

TANNER —
Prionus coriarius

The Tanner, like many large beetles, may vary considerably in size, ranging from 19—45 mm. long. The reason for this is chiefly found in the varying nutritious qualities of the rotting wood on which the larvae feeds. If this is high and there are no set-backs such as prolonged cold periods during the summer time, growth would be greater than if the wood and the climatic conditions are poor.

In most beetles there are no easily seen differences between the sexes. In this species, however, the antennae of the male are longer and considerably more strongly notched in appearance than those of the female.

The beetles are active during July and August and may sometimes be seen in daylight as they emerge from the tree stumps in which they have developed. Although found through much of Europe it occurs in southern Britain only and is totally unknown from Scotland or Ireland.

Ergates faber

This beetle, which is widespread in coniferous forests in Europe, is sometimes imported into Britain with cargoes of softwoods, but does not survive to breed here. It is the giant of the European longhorn beetles for a large individual may reach 60 mm. in length although the same controls on growth occur as with the Tanner so that small or stunted specimens may be no more than 23 mm. long.

The beetle illustrated opposite is clearly a female for in males the antennae exceed the length of the body. These beetles are to be found for a short time in midsummer, for the life of each adult is only about three weeks, in all and after mating and egg laying, which is normally done at night, they die.

In some places, where there is plenty of moist, rotten wood, *Ergates faber* may be very abundant and the eggs may be deposited on telegraph poles, fence posts or even at times on house timbers. However, in such places the larvae seldom develop satisfactorily, and their main habitat is old pine or spruce stumps where they live for several years, growing to a length of about 65 mm. before pupation.

Spondylis buprestoides

This is another European beetle of pine or spruce forests, very occasionally introduced into Britain with

Thanasimus formicarius

Prionus coriarius

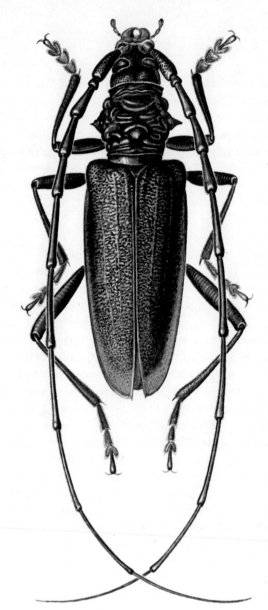

Cerambyx cerdo

Spondylis buprestoides

Ergates faber

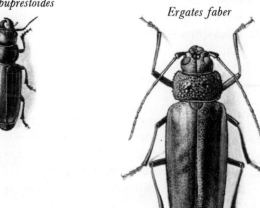

coniferous timber. It is much smaller than *Ergates faber* growing to about 22 mm. at most. The females lay their eggs on freshly cut stumps of trees and the larvae feed on wood at a much earlier stage of decomposition than is possible with many other forest species. The adults appear in late Summer but they hide throughout the day, only becoming active at twilight.

Cerambyx cerdo

In the broad-leafed forests of Europe, it is sometimes possible to see on the trunks of the trees oval holes measuring about 10 millimetres in one direction and 18 millimetres in the other. These are almost certainly the emergence holes made by *Cerambyx cerdo*, a handsome beetle associated particularly with oaks, but occasionally attacking other hardwood trees.

The remains of this beetle have been found in peat in Britain indicating that up to about 4,000 years ago when this country was heavily forested, its beetle fauna included some species now restricted to the Continent. Today it may be imported with timber, but does not survive there. The females, which may be up to 50 millimetres long, lay their eggs in small batches in the bark of the trees. These soon hatch, and the larvae feed in the first year on the bark and in their second year they begin to work their way into the sapwood of the tree. This often causes a characteristic discoloration of the tree trunk.

They continue to feed in this part of the tree for another two years, growing to a length of about 90 mm. They then bore for perhaps as much as 50 centimetres into the heart of the tree to pupate, emerging as an adult within about six weeks, but remaining in the safety of the pupal chamber throughout the Winter and finally appearing about the middle of the next Summer.

A severe infestation of these beetles may kill a number of trees within a given area, but it is not particularly common and does not normally cause such bad damage. If a tree containing pupae is felled for timber, the present method of kiln-drying does not destroy the insects. This being the case, the insects may be transported and may emerge some time later in an area far removed from their original home.

Rosalia alpina

This beetle, which may be seen flying in sunny weather in late Summer, is found almost exclusively in beech forests, mainly those in southern Europe standing at 600 metres or more above sea level. It occurs in Britain only as a rare import with timber. Although still common in some areas, its numbers have dwindled considerably in others, partly due to the destruction of the forests on which it depends, but also because of over-enthusiastic collecting by entomologists. This species with its blue and black striped livery is regarded by many people as one of Europe's most beautiful beetles.

Callidium violaceum

This beetle, which may grow to a length of 16 mm., is a beautiful metallic purple colour. It is widespread in Europe where the adults may be seen in early and middle Summer, but in Britain occurs only very sporadically.

The eggs of *Callidium violaceum* are laid on the bark of fallen or felled trees and the larvae feed on dry wood, boring deeply into the trunk to pupate. Any timber, especially softwood, in which the bark is not stripped from the wood is in constant danger of attack by this species. In Britain, ornamental rustic woodwork most often provides a home for it.

Leptura rubra

Leptura rubra, as with many insects, shows a size difference between the males and the females, the latter being the larger and growing to a length of up to 19 mm. The colour of the elytra varies also, those of the female being red, while those of her mate are yellowish.

The species is widespread in Europe and may be found in both lowland and mountain coniferous forests. Although imported into Britain it does not breed here. The adults are to be seen during the summer feeding on the blossoms of various plants which grow in the more open parts of the forest. The eggs are laid in stumps or wood piles or even sometimes telegraph poles near to such places and the larvae then develop in the damp wood.

Rhagium inquisitor

This beetle which is between 10 and 21 mm. long is found throughout most of Europe, including Britain, and also throughout North America. It is always associated with coniferous woods, where its larvae live beneath the bark of dead trees.

The adults emerge from their pupal stage late in the year but overwinter inside the tree, starting their active life as early as April in the following year. On a warm day they will fly fast and strongly and they have a jaunty, questioning look which has perhaps given them their name.

TIMBERMAN —
Acanthocinus aedilis

The Timberman is a beetle of pine forests, developing in rotting wood or occasionally under the bark of living trees. It is an easily recognised creature, for although the body is between 12 and 20 mm. long in the female her antennae are double this and in the male are even longer. The female carries a very distinctive ovipositor.

In Europe, including Scandinavia, the species is very common and widerspread. In Britain it occurs in two areas; these include a colony in Norfolk, which is almost certainly descended from introduced specimens, and a few colonies in the highlands of Scotland where they live in the remnants of the Caledonian forest which once covered much of the country.

Stenocorus cursor

This beetle, ranging in size from 25—32 mm. is found in many of the upland conifer woods in Europe and Western Asia, but is totally absent from Britain. There is considerable variation in colour within the species, the males usually being black and the females having red-striped elytra. *Stenocorus Cursor* is usually found in clearings either on tree stumps or

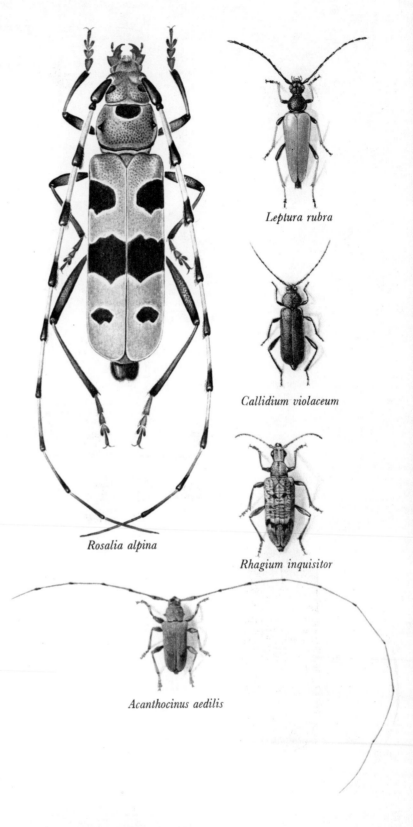

Leptura rubra

Callidium violaceum

Rosalia alpina

Rhagium inquisitor

Acanthocinus aedilis

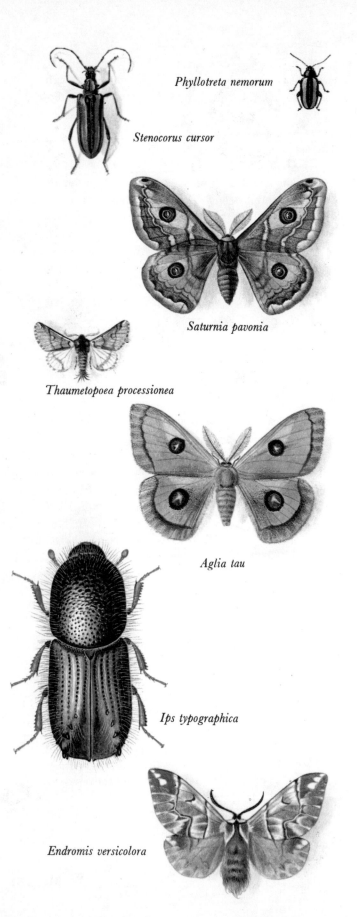

Phyllotreta nemorum

Stenocorus cursor

Saturnia pavonia

Thaumetopoea processionea

Aglia tau

Ips typographica

Endromis versicolora

on flowering plants and is active during most of the summer months.

FLEA BEETLE —
Phyllotreta nemorum

This little beetle, which measures only about 2 mm. in length is common throughout Britain, Europe and throughout Northern Asia into Korea. It lives on cruciferous plants (members of the wallflower family) in clearings and open places in woodlands.

The Flea Beetle's hind legs are very large and strong and it can hop a considerable distance when alarmed, a habit which has given it its name, for it is quite unrelated to true fleas. The larvae feed on leaves, reducing them to skeletons and sometimes killing the plant. The adults also eat leaves but they are less damaging.

Ips typographica

This beetle, which is only about 4 mm. long is none-theless one of the most serious pests of conifereous forests in Europe. A severe infestation can destroy mature trees though strangely in Britain it is rarely so damaging. The female bores a 'mother' passage beneath the bark of a living tree and lays about 60 eggs there. These hatch out within a fortnight and begin feeding on the inner bark, tunnelling at right angles to the original mother passage. Under favourable conditions the beetles complete their development within about a month and reproduce within a few weeks more, so that great damage can be done in a short time.

Hyphantria cunea

Hyphantria cunea ranks quite high among the pest animals imported into Europe from abroad. Originating in America where it is called the Fall Web-worm it is common in Central and Eastern Europe where three generations may easily be produced in a single year. Fortunately it has not yet spread to Western Europe or Britain.

The overwintering caterpillars pupate in early Spring and the adult moths may first be seen in deciduous woodlands in May. With a wingspan of 25 mm. they may be easily recognised by their white wings spotted with black and the lemon yellow abdomen.

The underside of leaves are chosen by females to lay large batches of eggs which hatch into hairy, brightly coloured caterpillars. These spin large communal webs and remain together for protection. They will eat the leaves of almost any deciduous tree, but frequently attack maples, poplars, willows and fruit trees. They completely defoliate the branch they are feeding on before moving on to the next and cause extensive damage wherever they occur.

OAK PROCESSIONARY MOTH —
Thaumetopoea processionea

This moth, which has a wingspan of about 30 mm. is on the wing in the tree tops in August and September but is attracted to light where it may most easily be seen. The female lays about 200 eggs in a batch which she disguises by covering them with hairs from her body. The caterpillars hatch next Spring and spin large silk shelters in which they live during the day, feeding only at night.

When moving from one tree to another the caterpillars travel in single file, a curious habit which has given them their name. Their bodies are covered with hairs which are intensely irritating to anything which they touch. Because of their defoliating activities they are major forest pests, and extreme measures are taken to try and control them include spraying from the air.

EMPEROR MOTH —
Saturnia pavonia

Emperor Moths, in which the wingspan of the female is about 75 mm. are found over much of Britain and most of Europe, frequenting heathlands and the edges of woods, where they may be seen flying strongly in late Spring. The caterpillars feed on a wide variety of plants including heather, sloe, willow, bramble and some herbaceous species as well. The pupa, protected by a dense coat of silk, has an ingenious 'door' at the narrow end through which the emerging moth can easily escape, but which prevents enemies from entering the cocoon.

Aglia tau

This strikingly coloured moth, which has a wingspan of about 65 mm. is found in the springtime in open beech forests in Europe but does not occur in Britain. The male has a rapid zig-zag flight, while the female, which is less active, remains close to the ground on the trunks of the trees. The green caterpillars, camouflaged by pale diagonal stripes on their sides feed on the beech leaves during the early summer months, but pupate by July.

KENTISH GLORY —
Endromis versicolora

This moth, which is now found over much of Central and Northern Europe also used to be common in southern England, but now occurs more in western Britain than elsewhere. The adults are to be seen in early Spring, the females at about 70 mm. wingspan being considerably larger than their male counterparts. In spite of this they are less active, not flying until dusk, while the males fly strongly in the sunshine. The eggs are usually laid on birch twigs, although some other broad-leafed trees may be food plants. At first the caterpillars are social, but as they grow they become more solitary, finally pupating in July.

PINE LAPPET —
Dendrolimus pini

Although there is an old record of this moth being captured in Britain, it cannot really be considered to be a native species, although it is common enough in mainland Europe. As its name suggests, it inhabits pine woods, but when resting on the trunks and branches of the trees the colouring of its wings makes it almost impossible to see.

The wingspan of the male is about 55 mm., and that of his mate is nearer 80 mm. but he is the more active of the pair. They may be easily distinguished by their antennae, which are comb-like in the male but like simple threads in the female. Their eggs are laid in clusters on the bark or needles of the trees and in late Summer the caterpillars emerge and start to feed. After three moults they start to descend to the ground to hibernate but as soon as it is warm enough (about 5° C.) in the next Spring they return to the tree tops to feed.

The caterpillars have mostly completed their development by midsummer, and then pupate, although a few hibernate for a second Winter. Although these caterpillars have many enemies among the caterpillar hunting beetles, for example the Tiger Beetles and Caterpillar Hunters (see page 20), sometimes great populations build up and cause severe damage to pinewoods.

OAK EGGAR —
Lasiocampa quercus

This moth which is widespread over Europe reaches up to 70 mm. wingspan in the female and occurs in a number of forms including one in Britain which lives in open heathland, rather than the oak forests which are their usual home. This subspecies has a wide food preference although it is rarely found in pine forests.

The typical form survives for one year, only the adults flying in July and August, while the other form takes two years to complete its development. The caterpillars hatching from eggs laid that year, hibernate and pupate next May in a cocoon in the ground. The caterpillars grow to a length of 70 mm. and are covered with brownish yellow hairs, which probably make them extremely unattractive to bird predators.

BLACK ARCHES —
Lymantria monacha

This moth gets its name from the pattern of black zig-zag marks on its forewings which provide it with a perfect camouflage when it is at rest on the tree trunks. There is a wide range of pattern and colour including a black form. The male which has a wingspan of up to 45 mm. has an abdomen which is square

in shape at the end, while the female, which is larger, always has a more pointed abdomen, the tip of which is bright red. The moth may be seen flying at night during July and August and is known from southern Britain but it occurs eastwards across the whole of Europe and Asia as far as Japan wherever the habitat is suitable.

The caterpillars which hatch in Spring from eggs laid the previous year may feed on a wide variety of leaves. It has been calculated that a single one may eat about thirteen hundred spruce needles in its nine weeks of larval life, so they may become a major forest pest since each female will lay several hundreds of eggs. The silk cocoon enclosing the pupa is spun into a crack in the bark. It is brown and hairy but has an almost metallic shine. The adult moths do not feed but have a short life which ends shortly after reproduction.

GIPSY MOTH —
Lymantria dispar

The Gipsy Moth, which occurs through most of temperate Europe and Asia used to be present in Britain, but any seen there now are likely to be the offspring of artificially reared specimens, for the species became rare and finally extinct here in the middle of the last century.

Strangely, when it was introduced into North America it multiplied to the extent of becoming a pest which has defied all attempts to eliminate it. The moths are on the wing in August. The male, which is a dark brown colour has a wingspan of about 40 mm. The female is white with pale zig-zag lines and a dark inverted 'V' on each of her forewings. She measures about 70 mm. in wingspan and has a stout, truncated looking abdomen and thread-like antennae.

The eggs are laid in batches on the bark of broad-leaved trees, and are covered with small scales from the female's body. They hatch during the next Spring and the caterpillars then feed until they are ready to pupate in late June, by which time they measure about 70 mm. in length. They feed chiefly on the leaves of oak, hornbeams and limes but many other trees, including fruit trees, may be seriously defoliated by them as well.

Lasiocampa quercus

Dendrolimus pini

BROWN TAIL —
Euproctis chrysorrhoea

Although fairly common in Central and Southern Europe the Brown Tail Moth is confined to the southern and eastern coastal areas of Britain. It has been accidentally imported into America where it has become one of the major pest species. The adult, which has a wingspan of about 35 mm., may be seen in Great Britain in late Summer on the leaves of various trees.

The batches of eggs laid by the female are camouflaged with hairs from her body and when the caterpillars finaly hatch they hibernate in a ruther communally spun web almost immediately. They make further webs as they feed and grow and several may even pupate together in the protection of the silk shield which they have spun. In Europe the main food plant of the Brown Tail is oak, and the caterpillars are forest pests. In Britain several plants are eaten and the numbers rarely get out of hand.

CLIFDEN NONPAREIL —
Catocala fraxini

This beautiful large moth, with a wingspan of up to 90 mm., is to be found in southern England and from

Lymantria monacha

Euproctis chrysorrhoea

Lymantria dispar

Scandinavia to Siberia on the Eurasian landmass. The caterpillars, which grow to a length of 100 mm., feed on poplars and aspens through the early summer months and the moths fly from late July into September. The moths are perfectly camouflaged as they sit on the trees which are their daytime resting place. If disturbed, the flash of lavender colour on their hindwings immediately catches the eye of the intruder but the moth then drops to the ground and with its wings folded runs to some place to hide among fallen leaves, where it will once again be completely camouflaged.

Catocala elocata

The Red Underwing, which is a southern English species, is extremely similar in appearance to this moth, but *C. elocata* occurs only in Central and Southern Europe and parts of Asia.

Although the wingspan is about 75 mm. *C. elocata* is so well camouflaged as it sits on the trunks of trees that it is very rarely seen by day. If frightened by anything it reveals the bright coloured hindwings, which surprise and help to scare away enemies such as small birds. The caterpillars may be seen in early Summer, feeding mainly on both poplar and willow leaves. The adults are on the wing from midsummer to Autumn.

LARGE EMERALD —
Geometra papilionaria

This striking green moth, which has a wingspan of about 45 mm., is common from June to August in open woods throughout the British Isles and Europe, except for the most northerly parts. The caterpillars which hatch in late Autumn hibernate on a silk mat which they spin on a twig of birch or one of their other food plants such as hazel or alder. They feed early during the next year and pupate on the ground by the summer months.

GARDEN TIGER —
Arctia caja

The Garden Tiger, which has a wingspan of about

70 mm., is to be found throughout Britain and across Europe except for the extreme south, flying in July and August. It is a beautiful but variable moth and it is likely that no two specimens are exactly alike. The caterpillars, which are covered with very long dark hairs, are known as woolly bears. These hatch before Winter when they hibernate, but after they awaken in the Spring they may often be seen on nettles, docks and a wide range of other plants on which they feed. They sometimes become a local pest, but they should be handled with care for the hairs can cause an unpleasant skin rash.

PINE HAWK —
Hyloicus pinastri

This inhabitant of pine forests occurs throughout Northern and Central Europe, but is rather uncommon in Britain, although its numbers seem to be increasing there. With a wingspan of 85 mm. it has the fast flight of other members of its family, but is normally active only at night, when it seeks out honeysuckle and other long-tubed flowers from which it feeds. During the daytime it sits on the pine trees perfectly camouflaged by the grey and brown shading of its wings.

The caterpillars, which are coloured green with pale stripes, are camouflaged as they sit among the pine needles on which they feed. Although Pine Hawks could easily become forest pests they rarely do so for their numbers are kept down by numerous species of woodland animals which parasitise or prey on them.

WOODLAND BROWN —
Lopinga achine

The Woodland Brown Butterfly may be seen on the wing in June or July in shady deciduous woods mainly in North Central Europe where scattered colonies occur. The caterpillars, which are green with one black and two white stripes down their sides, feed on various grasses including wheat. They hibernate in a fairly advanced stage, pupating quite soon after their reappearance in late Spring.

Catocala fraxini

Catocala elocata

Arctia caja

Hyloicus pinastri

Geometra papilionaria

Lopinga achine

Nymphalis antiopa

Limenitis camilla

CAMBERWELL BEAUTY —
Nymphalis antiopa

It is many years since the Camberwell Beauty occurred commonly in Britain. Nowadays only occasional individuals which have migrated from the Continent are found. While it is rarely common there, it is known from a great range of country, particularly where there are hills or mountains. It occurs from the south (except Spain) to the most northern pars of Scandinavia. It is found also in North America, where it is called the Mourning Cloak. It flies in open country early in the year when it first emerges from hibernation or late in the Summer. The caterpillars feed communally on such forest trees as willow or birch or elm but are never sufficiently abundant to become a pest to the trees.

WHITE ADMIRAL —
Limenitis camilla

The White Admiral, which has a wingspan of up to 60 mm., may be seen in the midsummer months from southern England and Scandinavia across Central Europe. It inhabits lowland deciduous woods, often resting on the flowers of blackberry or raspberry which provide it with nectar. The caterpillars feed on honeysuckle and hibernate in the Autumn when they are well grown, pupating after a further period of feeding during the springtime. As with most other members of its family, the structure of the front legs is the only clear way of identifying the sexes.

ST. MARK'S FLY —
Bibio marci

In the Spring swarms of St. Mark's Flies may be seen, their long legs dangling as they cruise slowly round in the sunshine. These black, hairy insects which are about 12 mm. long, are the males, for the females sit unobtrusively on flowers and vegetation beside paths and rides. They have short, rather stout-looking antennae, which are placed below the eyes. The adults can often be seen hovering over grassland. The eggs are laid in humus-rich soil, where the larvae, if very numerous, may damage tree crops.

Episyrphus balteata

This common, handsome fly which is about 8—11 mm.
long, may be found on flowers in woodland clearings
or by pathways during the late Summer and Autumn.
Its larvae feed on aphids, especially those which infest
cabbages, so it is a valuable insect to farmers and
gardeners.

Exorista larvarum

This fly, which is rather variable in size, ranging up to
15 mm. long, is to be found commonly in clearings
and sunny places in woods. Here the females search
out the caterpillars of various defoliating insects and
lay their eggs on them. The fly grubs, when they
hatch, bore their way into the body of the caterpillar
and feed on its tissues, but do not normally kill it until
it is ready to form a chrysalis, at which time the fly
larvae, fully developed themselves, leave their host's
body and crawl to the ground to pupate.
In Britain the Drinker Moth Caterpillar is among
the many insects parasitised, while in Europe the
Brown Tail and the Gipsy Moth are among the

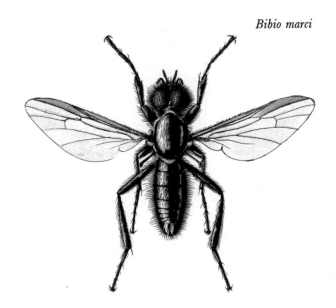

Bibio marci

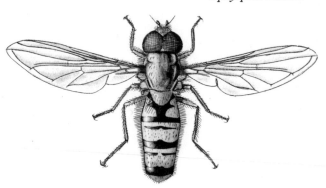

Episyrphus balteata

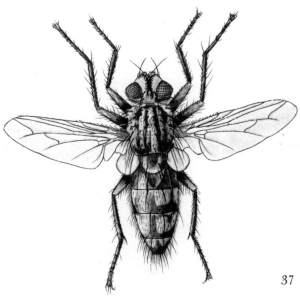

Exorista larvarum

Ornithomyia avicularia

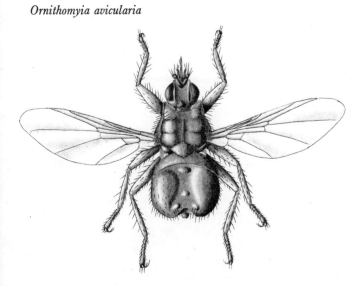

37

hosts. This fly has been especially imported into America as part of the campaign to control the Gipsy Moth (see page 32).

BIRD LOUSE FLY —
Ornithomyia avicularia

A common parasite of many European woodland birds, especially when in the nestling stage, is this Lousefly. Measuring 5—6 mm. long, it is fairly large compared with its host and a severely infested bird may be weakened by the loss of blood which it takes. Its large claws enable it to scuttle sideways through the plumage and help it to escape when the bird is preening or scratching and the long wings enable it to find another host if need be. The female produces fully developed larvae, which pupate as soon as they are born.

AMPHIBIANS

EUROPEAN SALAMANDER —
Salamandra salamandra

The vivid colours of the European Salamander gave rise to legends that this animal thrived in fire. However nothing could be further from the truth, for the Salamander, in common with all other amphibians, has a soft, moist skin and must be in a humid environment to survive. This species does not occur in Britain, but is widespread in Central and Southern Europe particularly in upland forest areas where there are clear streams. Here it may be found at altitudes of up to 1,200 metres above sea level. It generally lives near to water in moss or among stones or in small cavities in the stream bank, and normally becomes active in the evening or after rain, when it leaves its hiding place in search of food. It is attracted by the movement of small invertebrates and pursues earthworms, snails, spiders and insects which are sufficiently slow moving to be caught by it, for the Salamander cannot run fast in hunting its prey or in any other activity. This is the reason for the bright colour, for the Salamander cannot defend itself in any usual way. Instead it has glands in the skin which produce a highly unpleasant secretion which deters predators which might otherwise kill it. The colours are a warning that is quickly learnt and heeded by the hunters.

During the Winter Salamanders hibernate, individuals from a wide area often coming together and burrowing into the ground in the same ice-free place. Hibernation ends in about early April and soon after becoming active the females start to produce their young. These are born as tadpoles about 30 mm. long; have gills and four well developed legs. They normally take 2—4 months to complete their metamorphosis and soon after this they may leave the water, although they are not mature for another four years. Shortly after the birth of the young the adults mate again, for the development of the embryo takes about 10 months. Salamanders are long-lived animals and a life span of over eighteen years has been recorded.

COMMON FROG —
Rana temporaria

The Common Frog is a widespread species in Europe, including Britain, although it is now less numerous than it used to be. The reasons for this include the destruction of their breeding habitats which include ponds, ditches and small streams and the widespread use of insecticides, many of which are extremely poisonous to amphibians. It is still abundant in the more remote areas, including mountains up to 3,000 metres above sea level.

Soon after emergence from their winter hibernation, the frogs congregate at particular ponds or streams and it is at this time that the trilling song of the male frog may be heard. It is thought that they recognise their breeding places by the scent given by the special microscopic plants in the water which are the food for the tadpoles. The females lay between one and three thousand eggs in clumps in shallow water. The unprotected spawn soon hatches into fish-like tadpoles which develop into froglets in about two months, although occasionally they may take much longer. The adults move away from the water after breeding, but return to it again in the Autumn for hibernation, which is generally in the mud at the bottom of ponds. Frogs' food includes many sorts of insects and their larvae, spiders, slugs, snails and worms.

REPTILES

SAND LIZARD —
Lacerta agilis

Sand Lizards are found through much of Europe, south of Scandinavia. In Britain they are rare and live only in a few localities in dry heathlands or dunes; on the Continent they are found in hedgerows and in clearings in woods. They grow to a length of about 200 mm. and may be recognised by the three lines of dark patches with white spots on the backs and sides. Males tend to be brighter and greener in colour than the females, which are more grey. Sand Lizards often live in colonies, scraping out holes for themselves or occupying old mouse holes, usually on well-drained slopes. They love sunshine but cannot stand great heat and retreat to their holes or the shadow of vegetation during the hottest part of the day. They feed on many sorts of invertebrates, including grasshoppers, beetles, spiders, worms and slugs and occasionally even young lizards.

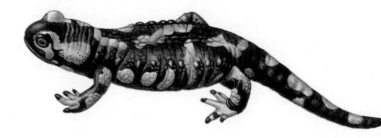

Salamandra salamandra

Rana temporaria

Lacerta agilis

Vipera berus

Anguis fragilis

During the breeding season in May and June there is a good deal of rivalry between the males, who, having chosen a mate stay close to her and repel all other suitors. Depending on her maturity the female lays between six and fifteen eggs in a small hole which she digs especially for them. Their time of hatching varies with the temperature: it may be as little as seven or as much as twelve weeks. Young Sand Lizards measure about 60 mm. at birth, but have increased their size by a third by the time that they hibernate. They are able to breed in their third year and are probably full grown by the time that they are four or five years old. Hibernation in Britain is from late September to April, probably in the same holes in which the animals pass the Summer. Sand Lizards have many enemies. One defence which they share with many other lizards is the ability to lose the end of the tail, and then subsequently growing another.

SLOW-WORM —
Anguis fragilis

Slow-worms are often mistaken for snakes, since they have no legs, but they are in fact legless lizards. Growing to a length of about 500 mm. they are extremely variable in colour, including browns, greys, and dull reds on the upper parts and grey below. The scales are tiny and give the animal a polished appearance. The name Slow-worm is quite appropriate for although it can move fast if frightened, it does not normally do so. Nonetheless the species is widespread, being found from central Scandinavia southwards and eastwards into the Middle East in a very wide range of habitats, where there is some cover in which it can shelter.

The Slow-worm's food consists largely of many types of invertebrate, but it eats the small white slug *Agrolimax agrestis* in preference to anything else. Mating takes place in late Spring and early Summer and the young are normally born late in the Summer. The species is ovoviviparous which is to say that the young are both retained and protected inside their mother until they are ready to hatch, but there is no nourishment given to them from the maternal tissues as in the case of mammals. The number in a litter depends entirely on the age and size of the mother. Slow-worms hibernate, often communally, from late October until March when they emerge and may be seen basking in the early spring sunshine.

ADDER OR VIPER —
Vipera berus

The commonest British snake is the Adder, which is to be found in a variety of habitats, preferring a dry environment such as heathland or a wood with open glades in which it can bask. In mainland Europe it is the most widespread of the snakes, occurring from northern Scandinavia southwards and absent only from the hottest parts of the Continent.

It is recognisable chiefly by its rather thickset shape (about 800 mm. long at most), and its skin which is decorated with a zig-zag pattern of dark marks down its back that are less distinct in females. The ground colour varies from cream to grey or olive green and in a few cases the animal may even be entirely black.

Adders feed on large invertebrates and small vertebrates too, including amphibians, mice, young birds and particularly lizards. These are all killed by the venom injected by the Adder as it bites and this is quickly fatal to them. If the prey animal is not killed instantly by the poison, the snake tracks it down, using its forked tongue to taste the scent of the prey. Adders bite large animals such as dogs or humans purely in self defence; although extremely unpleasant the effect of the bite is very rarely serious but medical aid should be sought as quickly as possible if a person is bitten.

Adders mate in late Spring which is the time that the territorial display called the Adders' dance may be seen. In this two male Adders, facing each other, raise the fore-parts of their bodies to try and force each other to the ground. The loser is unhurt but rushes away. The female produces between six and twenty young in late Summer. In Autumn, when the temperature falls below 8° C. the Adders go into hibernation, often singly, but sometimes several together in a hole in the ground from which they emerge early in the next Spring.

BIRDS

RAVEN —
Corvus corax

The deep-throated call of the Raven is often the first indication of the bird's presence over much of the western and northern parts of Britain and the rest of Europe other than France and Italy. This largest member of the crow family which measures over 630 mm. long, may be seen soaring on large square-ended wings, with its spread primaries looking like fingers.

During their courtship the birds perform aerial acrobatics, diving, soaring to great heights and even turning upside down in flight. Perhaps because they are large, their numbers have been greatly reduced in many areas, although they now seem to be increasing slightly again. Where they are plentiful, or where suitable food is abundant they may congregate in large flocks. But they are normally single birds, and pairs or family groups is all that are seen.

The Raven builds its nest in the upper branches of a large tree, or on inaccesible cliffs including those overlooking the sea. The same site is utilised over a number of years and the nest is added to and refurnished each season. It is in any case a bulky affair with a base of large twigs lined with grass and hair. The four to six eggs may be laid as early as February: they are light blue in colour, speckled with dark brown and are incubated by the female for the greater part of three weeks, before the chicks hatch, although she is sometimes relieved by the male. The young birds remain in the nest for another six weeks before they are fledged and then stay with their parents until early in the following year.

Ravens feed on a wide variety of animal food, but kill only small creatures and insects. Carrion forms an important part of their diet ranging from the placentas of sheep in hill country to stranded whales on the sea shore.

Ravens and their relatives are among the more intelligent of birds. If one is reared by humans, the bird will decide where in the family 'peck order' it belongs, and may attack domestic animals and even children which it considers to be below it socially.

Corvus corax

Corvus corone corone

Corvus corone cornix

41

CARRION CROW —
Corvus corone

Although the Carrion Crow is also a black bird it may be distinguished from the Raven by its smaller size (about 460 mm. long). Another factor is that it rarely soars, while its call is a croaking caw usually repeated several times rather than the wide vocabulary of the Raven or Rook. Another bird that is very closely related is the Hooded Crow *(Corvus corone cornix)*. This replaces the Carrion Crow in Ireland, northern Scotland and eastern Europe and is easily recognised by its grey back and under-parts. One or other form is to be found throughout Europe and along the line which divides them there is some interbreeding.

Crows feed on a wide variety of food including small mammals and birds, carrion, seeds and invertebrates including molluscs which they drop onto roads or rocks to break the shell. If they have any surplus food they bury it. They are regarded as enemies by farmers and gamekeepers, however, for they may kill weakly lambs and eat game birds as well as their eggs. In spite of this they are not uncommon, living in a wide variety of habitats, preferably where there are some large trees in which they breed. Recently persecution has probably been less than in the early years of this century and the Crows have become more abundant, even making a living in large cities like London.

Although generally solitary they may be gregarious at roosts, and during the winter months, when flocks of birds from the far north migrate into Central Europe. Their nests are large affairs of sticks usually built high in tall trees in which the female incubates the five or six eggs for three weeks and where the young must remain for another month before they are fledged.

JAY —
Garrulus glandarius

Jays are common in deciduous and mixed woodland throughout Europe with the exception of the north of Scotland and Scandinavia. The most handsome of the crow family in Europe, there are slight differences between the typical European birds and those found in Britain, which are on the whole smaller and pinker in colour although they are similar in habits and behaviour.

They are extremely wary birds, alert to danger at all times; it is the raucous alarm call of the Jay which warns the other forest birds when a human being or any potentially dangerous animal shows itself. The nest which is usually made in dense vegetation, is a fairly substantial nursery made of twigs, fine roots and moss or hair. The five to seven greenish grey eggs, speckled with dark brown are laid in late April or early May. Both parents share the task of incubation which lasts about sixteen days and both feed the chicks until they are fledged in another twenty days.

The young birds remain with their parents until the Autumn, when migration takes place among birds in the colder parts of Europe. Jays do not travel far at this time, merely moving to areas where the ground will not be frozen throughout the winter, a good many coming to Britain, where the species is residential. Jays feed on plants and animals including many kinds of nuts, berries and shoots, small mammals and amphibians and the eggs and young of other birds, as well as snails, worms and insects. In the Autumn they may hoard nuts and acorns but often forget the site of their larder so the acorns germinate and grow during the next Spring. Jays are thus important agents of regeneration in the forest.

GOLDEN ORIOLE —
Oriolus oriolus

Another cautious and timid bird of dense forest cover is the Golden Oriole. From May onwards it gives its presence away with its beautiful fluting song, or its harsh screaming alarm notes. It is found almost throughout Europe except for Scandinavia and southern Greece, but few reach England, where it normally occurs in very small numbers in the southern counties. The hammock-like nest which is built mainly by the female, hangs from a forked branch. The eggs are laid in late May and hatch in about fourteen days; the young require another fortnight of care by both parents before they are ready to leave

Garrulus glandarius

Oriolus oriolus

the nest. The food of Golden Orioles varies according to the season. Insects, which are abundant in late Spring are eaten then. In Summer many kinds of soft fruit such as cherries, grapes, figs and berries are the mainstay of their diet while spiders and snails may be taken at any time. After rearing one brood the Orioles depart for South Africa and Madagascar where they spend the winter months.

NUTHATCH —
Sitta europaea

The British Nuthatch is an elegant bird, about 140 mm. long, blue-grey above with pale apricot colour below. Several subspecies have been described from mainland Europe, differing chiefly in the intensity of the colouring of their underparts, but they are similar in other respects. They are birds of mature deciduous woodlands, where they feed on many sorts of nuts and insects. Any nut which cannot be broken easily is wedged into a crevice in the bark of a tree and hamme-

Sitta europaea

Carduelis spinus

Coccothraustes coccothraustes

Loxia curvirostra

Pyrrhula pyrrhula

Certhia familiaris

red by the bird's strong bill. Insects are picked off the bark of trees by the birds which climb acrobatically up and down the trunks and branches without using their tails for suppport in the way that Woodpeckers do. The nest is made inside a small hole in a tree, and the entrance plastered up so that no larger birds can get in. The eggs, white with splashes of red-brown, are laid in early May and incubated by the hen for about sixteen days. The young are fed by both parents for over three weeks before they are able to fly.

HAWFINCH —
Coccothraustes coccothraustes

The Hawfinch, which occurs from southern Britain and Spain across to Eastern Europe, is the largest of the finches but is rarely seen even where it is fairly common for it is extremely shy and secretive, spending a good deal of its time among the foliage of the treetops. It nearly always lives in deciduous woods. The main identification features are its portly appearance and the very heavy conical bill which is obvious even in flight. This is powerful enough to crack the stones of cherries and even olives to extract the kernal. The nest is usually in a small tree often close to the trunk; sometimes several will be found together in adjacent trees. The eggs are laid in early May and hatch in about ten days. The young are fed chiefly on insects and are fledged within another twelve days. In Winter Hawfinches form flocks; those in the colder areas move to parts of the Continent where the weather is less harsh, but over most of Europe the species is sedentary.

SISKIN —
Carduelis spinus

Spruce forests are the favourite home of the Siskin, which is a yellow-green finch about 114 mm. long, distinguished from all similar birds by the bright yellow flash on either side of the tail and by the dark head of the male. In the wintertime bands of Siskins move from their woodland breeding grounds and may be seen widely over much of lowland Europe, especially in Alder groves, where there are plenty of seeds for them. In Britain the Siskin seems to be becoming

more common and in recent years has appeared in unprecedented numbers in the southern counties, probably because of a general fall in temperature. The nest is usually sited at the end of a branch of a tall conifer. Here the female lays four to six pale blue eggs, marked with red-brown blotches, which she incubates for about twelve days. The young remain in the nest for about fifteen days, fed chiefly on insects by both parents until they are able to fly.

BULLFINCH —
Pyrrhula pyrrhula

The Bullfinch, which is a plump bird 145 mm. long with a very stubby beak has a black head, bright pink breast and grey back with a white patch over the rump. It is primarily a bird of forests but has adapted readily to hedgerows and gardens where it is often seen. Here they are not always welcome for as well as fruits and seeds, Bullfinches have taken to eating buds of fruit trees and ornamental plants which they can damage severely. The nest is a deep cup of grass, hair and moss, securely hidden in a hedge or low dense vegetation. The eggs are laid in late April and hatch after about twelve days incubation by the female. The young remain in the nest, being fed on insects and seeds for another sixteen days.

CROSSBILL —
Loxia curvirostra

Crossbills are birds of pine forests and are found in southern Scandinavia, most of Eastern and Central Europe and Spain. The only regular breeding areas in Britain are in northern Scotland and East Anglia but they are sometimes seen beyond these limits

Anthus trivialis

Parus ater

Parus cristatus

Regulus regulus

Parus palustris

for the species is one in which occasional population increases occur and for a while they may nest in places which are only marginally suitable. The breeding success of the Crossbills depends on the crop of conifer seeds, for they eat nothing else. The crossed bill which is not always noticeable, is adapted to twist the bracts of the cones of spruce and pine so that the seeds may be extracted. The presence of cones broken open in this way is a sure indication of Crossbills in the area. The call is the unmistakeable and deliberate 'kip, kip, kip.'

Breeding is rather less seasonally regular than with many other birds, although the eggs are usually laid during the first four months of the year. The nest, built by the female alone, is generally high in a tree on the edge of a wood. Three or four eggs are laid and incubated by the female who is fed by her mate. The chicks hatch in about thirteen days and are cared for for a prolonged period by their parents, as they cannot fly for several weeks. During this time their bills are straight and they do not begin to be crossed until the young are independent and feeding for themselves.

TREE CREEPER —
Certhia familiaris

These tiny, agile birds may be seen on tree trunks looking for the insects and spiders which are their food. They normally move head uppermost in a spiral, and fly down to a lower point to start a fresh search. They occur in woodlands and large gardens throughout Britain, Central and Eastern Europe, including southern Scandinavia. A habit which they have developed recently is to excavate roost-holes in the fibrous bark of sequoias, which are large conifers introduced from America during the last century. The discovery of these little hollows, about 30 mm. in diameter is an indication of Tree Creepers. The nest is unobtrusively hidden behind loose bark, or cracks in tree stumps. The eggs are laid in late April and incubated by both parents. Two broods may be reared and in the wintertime family parties may join up into small flocks working the trees together. The call is quite simply a single, very high note.

TREE PIPIT —
Anthus trivialis

A bird which may be seen throughout Europe except for the most southerly parts, Ireland and the wildest parts of Scandinavia is the Tree Pipit. About 150 mm. long, it lives in areas of scattered trees or clearings in woods where there is dense ground cover in which it may hide its nest. It is a summer migrant, arriving at its breeding haunts in late Spring and from then on the trilling song of the male may be heard as he glides down with wings outstretched in a flight from tree top level to the ground. The eggs are very variable in colour, and are incubated by the hen bird alone, but both parents feed the chicks after they hatch and continue to care for them even when they have left the nest. The Tree Pipit feeds on many kinds of insects, especially weevils, also on spiders and other small invertebrates. The southwards migration is delayed until the end of the Summer and Tree Pipits may often still be seen in Europe as late as early October. The call note is the characteristic 'teezee' sound.

CRESTED TIT —
Parus cristatus

The Crested Tit is another inhabitant of coniferous and mixed forests, but is found less widely in Europe than is the Coal Tit. It is one of the rarer of British species and is known only from northern Scotland where it feeds on insects and conifer seeds. Its large size (about 113 mm. long) means that it occupies a slightly different niche in the environment and feeds mainly on the trunks of the trees. The nest is usually made in holes, or cracks in tree stumps, although nest boxes may also be used .On the Continent the species is often double-brooded, but this habit is rare in Britain. It is less sociable than most tits.

COAL TIT —
Parus ater

Easily recognised by the white patch on the back of its head, is the Coal Tit which, measuring only 110 mm. long is one of the smallest members of the

Titmouse family. It is to be found through most of Europe from southern Scandinavia southwards favouring coniferous woodlands, but it often lives in parks and gardens, where it visits bird tables and shows little fear of human proximity. Like all members of its family, the Coal Tit is highly acrobatic and can cling to the smallest twigs of the trees while searching out the insects and spiders which are its summer food. In the wintertime it joins flocks of other small birds to forage through the woodlands and it feeds mainly on the seeds of conifers. As with most Titmice, the nest is made in a secure hole and two broods may be reared during the summer months. Its call is more plaintive than that of other tits.

MARSH TIT —
Parus palustris

An inhabitant of deciduous woods, the Marsh Tit is similar to the Coal Tit. It measures about 114 mm. long, and so is slightly larger than the Coal Tit. Its greatest similarity, however, is with its close relative the Willow Tit, from which it differs mainly in the glossy

rather than dull black feathers on the top of its head, and its preference for a slightly drier habitat. It rests mainly in holes in willow or alder trees, usually rearing only one brood a year. It feeds on insects and spiders eked out by seeds during the winter months when small flocks may roam the woods and even visit garden bird tables for food.

GOLDCREST —
Regulus regulus

The smallest European bird, measuring less than 90 mm. long, the Goldcrest is widespread but infrequently seen, for it occupies the tops of the trees in coniferous forests. Since it is generally green, it is easily distinguished from other tits. However, its high-pitched call may give its presence away for it is rarely silent when seeking the aphids, small beetles, spiders etc., on which it feeds. The nest is built hammock fashion, hung from a branch, and is made from plant fibres, lichens and spiders' webs warmly lined with hair and feathers. The female incubates the eggs alone but both parents feed the nestlings. A second brood is often produced in midsummer. In the wintertime Goldcrests often form small flocks searching communally for their food.

HEDGE SPARROW or DUNNOCK —
Prunella modularis

The Dunnock, which is no relation to the true sparrow, is a bird of woodlands, thickets and gardens, where its retiring ways often cause it to be overlooked. However, its sweet song may be heard in springtime, through much of Britain and Europe. It feeds mainly on insects and small seeds taken from the ground and when visiting garden bird tables normally takes fragments knocked down by other birds. The moss-lined nest is usually built in thick cover. The four to seven eggs are a clear blue colour and are incubated by the female alone. In Britain, Dunnocks are one of the species most frequently parasitised by cuckoos but this is not the case in Europe. British Dunnocks are residents; those in the more northerly parts of Europe migrate to the Mediterranean region for the winter months.

Prunella modularis

Muscicapa albicollis

Hippolais icterina

Troglodytes troglodytes

WREN —
Troglodytes troglodytes

Wrens live throughout Europe wherever there is dense cover in woodlands, scrub, river banks and even gardens. Although tiny (just over 100 mm. long) they are easily recognised by their quick flight from one perch to another and when at rest, the vertical position of the tail. For so small a bird the Wren has an outsize song — a loud scolding trill which may be heard from early Spring onwards as the male defines his territory and tries to attract a mate. Within the territory he will begin to build a number of nests, known as cock nests. These are little more than platforms of grass and moss, one of which will be selected by the female and completed to an oval shaped nursery entered at the side, where the young are reared. Wrens are usually double-brooded, rearing between four and seven chicks in each clutch. In wintertime they often roost together for warmth in any small crevice or hole that they can squeeze into. As many as fifty have been recorded using a single nest box in this way.

COLLARED FLYCATCHER —
Muscicapa albicollis

This bird, which is an inhabitant of South and Eastern

Europe, is an extremely rare visitor to Britain where very few have been reliably recorded. In size (about 127 mm. long) and appearance it is similar to the Pied Flycatcher, except that a collar of white surrounds the neck dividing the black cap from the dark back, and the patches of white in the wings are stronger than in the Pied Flycatcher. It nests in holes and hollow tree stumps and sometimes in nest boxes. Four to seven eggs are laid in May and after rearing one brood which remains with its parents for a while in the northern woods, they migrate to Africa for the winter months.

PIED FLYCATCHER —
Muscicapa hypoleuca

Towards the end of April the Pied Flycatcher returns from Africa to its breeding grounds in deciduous woods, often near water, in Wales and northern England. In mainland Europe it prefers coniferous forests as it does in Asia and North Africa. Although

49

Erithacus rubecula

Phylloscopus sibilatrix

Turdus viscivorus

not evenly distributed it is often locally common in some areas. Nesting in holes, it rears between four and eight young which may be seen, later in the season flying with their parents. All sorts of insects are eaten, some caught on the wing and others taken from the ground and vegetation. It does not return to the same perch after each capture as does its cousin the Spotted Flycatcher. A good identification characteristic is the movement of the tail which is constantly jerked and flirted as the bird moves about on the ground.

ICTERINE WARBLER —
Hippolais icterina

Found in Europe from central France eastwards, and occurring in Britain only as a very rare visitor, the Icterine Warbler is an inhabitant of woodlands, cultivated land and gardens throughout the summer months but returns to Africa during the wintertime. As with many of the Warblers the best indication of its presence is its song, which is a sustained jumble of sweet and discordant notes, each repeated several times. If a sight can be glimpsed of the singing bird the open bill will show bright orange. Because of the habit it has of erecting the feathers of the top of the head this often looks large and rounded. West of the area in which the Icterine Warbler is found, another species, indistinguishable in the field occurs. This is the Melodious Warbler whose song is sweeter and less repetitious. The two species overlap in Italy, where it might be possible to confuse them.

WILLOW WARBLER —
Phylloscopus trochilus

Willow Warblers, although only about 105 mm. long, migrate from Africa to the most northerly areas of Europe, reaching the British Isles in late March and early April. They are to be found in open woods, both deciduous and coniferous, where there is a heavy undergrowth. They may be confused with their close relative the Chiff Chaff, but may be distinguished by their songs, which in the Willow Warbler end on a series of descending notes, while the Chiff Chaff merely repeats its name over and over again from its song post. The colour of the legs is usually light brown in Willow Warblers where as those of the Chiff Chaff are very dark. Both birds feed on small insects, but are ecologically separated, where they occur together, by the preference of the Willow Warbler for low shrubs and even dense ground cover as a place to build its domed nest. The Chiff Chaff, meanwhile, takes a site higher up in tall shrubs or trees.

WOOD WARBLER —
Phylloscopus sibilatrix

Like most members of the warbler family, the Wood Warbler is more easily heard than seen, for the loud, liquid trilling of the male bird is performed in flight, as it moves through the trees. It may be recognised by its small size (it is about 125 mm. long), its bright yellow throat and white underparts which distinguish it from the Icterine and Melodious Warblers with whose range it overlaps, except in Italy. In Britain it may be found in deciduous woods throughout the area apart from Ireland and northernmost Scotland. Its domed nest is built on the ground, and the six to eight eggs incubated for thirteen days. Only one brood is reared in the season and in September the Wood Warblers migrate to Central Africa where they pass the winter months.

MISTLE THRUSH —
Turdus viscivorus

The Mistle Thrush, which measures 280 mm. long is the largest of the European thrushes. It is found throughout Europe except for Norway, inhabiting woods, orchards and gardens. Over the northern parts of its range it migrates southwards in hard weather; in Britain and further south it is sedentary. Although they form small flocks in the Winter, breeding starts early in the year and the loud cheerful territorial song of the male may be heard as early as February. This habit has given the bird the name of the Storm Cock in Britain where it is normally much tamer than in mainland Europe, where it is considered good eating. The nest which is built in the fork of a tree is a solid construction of mud

and grass. The eggs are incubated by the female, who is fed by her mate and the young, reared by both parents, leave the nest before they are properly fledged. Baby thrushes are often 'rescued' by well intentioned people who attempt to rear them. It is better to leave them to the care of their parents, who feed and guard them at this helpless stage.

ROBIN —
Erithacus rubecula

The Robin is to be found throughout most of Europe in gardens, parks and woods with thick undergrowth. In Winter it migrates from the colder areas but in Britain it is sedentary. In Europe it is generally regarded as a shy and retiring bird; in Britain it is quite the reverse; probably its confiding nature has made it the major factor in being voted the British National Bird. The melodious song of the Robin in Spring is an aggressive territorial defence; few birds

Picus viridis

Dryocopus martius

Caprimulgus europaeus

Dendrocopos major

are more bellicose at this time of year. The neatly
constructed nest may be in a thick hedge or a cranny
of some kind, but often a shelter for the nest will
be found in an old kettle or other such man-made
object thrown into a hedge. Young Robins do not
resemble their parents as they have speckled breasts
at first and soon after the first brood has left the nest
the parent birds rear another family.

GREEN WOODPECKER —
Picus viridis

Woodpeckers, as their name suggests, are mainly
arboreal birds, spending most of their time climbing
on tree trunks and branches, supported by their
short, stiff tails. They peck at the wood or trees infected
by insects or their grubs, such as those described

52

on pages 14 to 19, and tunnel into the larval burrows with their strong chisel-like beaks. They have extremely long tongues, which are sticky or harpoon-like and can be protruded several centimetres beyond the end of their beaks to capture the insects which they have exposed. Their flight is strong and rapid but undulating, with freewheeling glides alternating with wing beats. Most of Europe's nine native Woodpeckers are black and white birds, with flashes of red on their head or flanks. The Green Woodpecker, however, is largely olive green in colour although in flight the bright yellow rump patch is obvious.

The bird is to be found in deciduous woods over most of Europe other than the far north, Ireland and Scotland, making its presence known by its loud laughing cry, which has given it the country name of the Yaffle in parts of Britain. The nest is made in a hole bored in a decaying tree. As with all Woodpeckers, the eggs are a glossy white and the naked and helpless young are cared for by both parents. More often than other species Green Woodpeckers descend to the ground where they tear open the nests of ants which are one of their major foods. They may also be seen hopping around clumsily or perching across branches.

BLACK WOODPECKER —
Dryocopus martius

This bird, which is over 450 mm. in length is the largest of the European species of woodpecker. It does not occur in Britain for it is mainly found in coniferous forests and beech woods of northern and eastern parts of Europe.

The nest hole, which may be made quite high up a tree trunk, is usually started at a place where decay has set in, but as the nest is very large, often measuring more than half a metre deep, it is cut into still healthy wood. Both male and female work at the task of excavation which takes between two and three weeks. No nest material is used, but there may be a layer of wood chips on which the white eggs are laid. Both parents incubate and care for the young but once these are able to fend for themselves they are no longer tolerated by their parents.

Bubo bubo

GREAT SPOTTED WOODPECKER —
Dendrocopos major

Widespread in deciduous and coniferous woods the Great Spotted Woodpecker is one of the commonest members of its family through almost the whole of Europe. In springtime it is best detected by its drumming noise made by tapping rapidly with its beak on resonant dead branches. This sound which carries for a long distance through the forest, performs the same function as a song, in that through it the woodpeckers declare their territorial boundaries and attract mates. Almost all woodpeckers have this habit, but the drumming of the Great Spotted is louder and more frequent than most.

The nest hole of the Great Spotted Woodpecker is excavated in a tree trunk and the eggs laid by the beginning of May. As well as insects and their grubs and spiders, Great Spotted Woodpeckers feed on nuts to a large extent and during the Autumn and Winter they depend on them entirely. Hazel nuts are wedged into crevices in bark and hammered open and pine cones are treated similarly. Although they are very shy birds they may even come to feeding stations in gardens to take advantage of nuts and other foods there.

NIGHTJAR —
Caprimulgus europaeus

Nightjars, which are summer visitors to most of Europe except the extreme north, are to be found in open woodland and on commons and moors but are far less frequently seen than heard. As their name suggests they are active at night, which is when their curious churring song which may last for minutes on end can be heard. They hunt large insects, especially night flying beetles and moths, which they capture by sight, snapping them up with their very large mouths. They often swoop round domestic animals in the fields, and in several European languages the Nightjar has a country name which refers to the superstition that it takes milk from them. In England the name Goatsucker perpetuates this quite erroneous belief.

During the daytime the Nightjar is perfectly camouflaged in its brown and grey feathers and sits motionless either on the ground or lengthways along a branch, looking like a piece of dead wood. During flight the bird is silent, but it claps its wings together during display. The eggs of the Nightjar are laid in a scrape on the bare ground and are white lightly blotched with grey and brown and look so much like pebbles that they would normally be ignored by all but the most alert passers-by.

EAGLE OWL —
Bubo bubo

The largest of European owls, the Eagle Owl measures up to 690 mm. long. It is very rarely seen in Britain or most of northern France and the low countries, but may be found in suitable territory elsewhere over the Continent except for northern Scandinavia. The breeding site of the Eagle Owl is most likely to be among a tumble of rocks in forest, or on a mountain crag in more open surroundings; but it may also be in a hole among boulders on the ground or in a hollow tree or even take the deserted nest of a bird of prey. No nest as such is made but in late March or early April the female lays two to four eggs in the chosen place. These, like the eggs of all other owls are white and nearly round. Incubation starts with the production of the first egg and lasts about thirty five days. Since each of the eggs requires an equal length of incubation, the first laid hatches first and begins to grow before the next nestling is out of the egg.

Any owl family consists of birds of different sizes, for the oldest chick is able to beg more strenuously or grab more quickly any food that is available. The young, which are clad in a pale buff-coloured down, are fed by both parents, who in the early stages tear up the food in portions small enough for the baby birds to swallow easily, although owls have an extremely wide gape and can easily swallow large morsels. Small animals such as mice will be swallowed whole and the owl will later regurgitate sausage-shaped pellets containing the indigestable parts, such as the bones and fur. Eagle Owls hunt at dusk and at first daylight. They take prey up to the size of hares or capercaillie, (see page 62), and even Roe Deer, but most of the animals which they catch are smaller than this. The voice of the Eagle Owl is variable and includes the boo-hoo sound of the song as well as a number of other sounds such as bill clacking and spitting noises. Stories of ghosts and hobgoblins which abound in many remote parts of Europe may often have been started by travellers terrified by the cries of the Eagle Owl, heard at close quarters.

LONG EARED OWL —
Asio otus

The Long Eared Owl is essentially a creature of coniferous forests, and although it may occasionally be found in small areas of trees or in deciduous woodlands, neither of these habitats is normal for it. Within these habitat limits its range includes most of Europe except for the far north and the western part of the Iberian Peninsular. Although about 355 mm. long, it is not an easy bird to observe, even in those areas where it is quite common, for it is more strictly nocturnal in its habits than many of the other owls. If it is seen by daylight it is likely to be sitting bolt upright, in a curious elongated position close to the main trunk of a tree, or in dense foliage well camouflaged by its brown and grey colouring.

However, the position of the roost may often be found by the accumulation of pellets formed by indigestable material which the owl ejects soon after swallowing its food. Long Eared Owls often hunt beyond the confines of their forest home; here they

Asio otus

Strix aluco

may catch voles and mice from open ground as well as those species which inhabit woodlands. A number of birds have also been recorded among their prey, including jays and finches from the woods and sparrows and larks from farmland. Insects, especially the large forest beetles, form an important part of their diet in the summertime, and in times of plagues of cockchafers which occasionally devastate areas of mainland Europe, the Long Eared Owl feeds largely on these and is an important control of such pests. The long ear tufts of this owl, which stand erect if the owl is alarmed, have nothing to do with its power of hearing, but are probably associated with display behaviour in the breeding season. Other displays include zig-zag flights through the trees and a slow flight above canopy level, in which both birds of a pair clap their wings on each downstroke producing a loud cracking noise. They do not make a nest but normally take the old nest of a crow, magpie or even a heron or the drey of a squirrel. If no such place is

available the eggs will be laid on the ground, usually at the base of a tree. As with other owls, incubation starts with the first egg, and the older chicks stand a better chance of survival than the others. Both parents help to rear the young, which after an incubation period of about four weeks take nearly as long as that again before they are ready to leave the nest. Outside the breeding season, when the Long Eared Owl produces a wide variety of sounds, it is a silent bird, and not likely to alarm people with its call.

TAWNY OWL —
Strix aluco

The Tawny Owl is the most familiar owl in Britain although it does not occur in the extreme north

Cuculus canorus

of Scotland or in Ireland. Although it was originally a bird of forests, it is now to be found in parks and gardens and even in suburbs and towns. In Europe it is to be found across most of the Continent south of the latitude of southern Scandinavia. It is seldom seen, for it is nocturnal, but features which distinguish it from other owls are its lack of 'ears' and its thickset rather large-headed look.

The calls of the Tawny Owl include the familiar whistling hoot and the ke-wick call which it makes when hunting. Its food, which is caught at night, includes small rodents, shrews, small birds and insects. Usually the prey is sighted or heard on the ground, for like all owls, the Tawny has excellent eyesight and powers of hearing. The bird drops silently and catches its prey in its talons. Silent flight, which is an attribute of owls in general, is possible because of the softness of the plumage, which muffles the sound of wingbeats. Daytime and night-time resting places tend to be different, but may be detected by the digestive pellets found on the ground below them. If small birds discover an owl during the daytime they may band together and mob it, screaming and fluttering near to it, but not daring to approach too close for fear of its talons. Eventually the owl may be driven off to another resting place by the hullaballoo and the little birds resume their normal behaviour. At night, however, the owl may retaliate and has been recorded as beating the tree in which roosting small birds are sitting with its wings and driving them into the open where it catches them. Tawny Owls usually nest in hollow trees or the old nests of large birds

such as crows, and sometimes on the ground in rabbit burrows and occasionally inside buildings. The female incubates the three to six eggs for four weeks and the nestlings then have a long period of dependence before they leave the area in which their parents live.

CUCKOO —
Cuculus canorus

Relatively few people have ever seen a Cuckoo but most are familiar with the simple repetive song of the male which rings through woodlands and bushy ground during the early Summer, although the 'bubbling water' call of the female is less well known. The Cuckoo is one of the most widespread of birds, occurring from Ireland across Eurasia to Japan and going as far north as the Arctic Circle. The Cuckoo is the only truly parasitic bird in Europe, for the female lays her eggs in the nests of other birds, which act as foster parents. The choice of host is not haphazard, for she usually looks for the nest of the same species as the one in which she was reared. When the laying of eggs in this nest is nearly complete, the Cuckoo adds one of hers and removes one of the host's eggs, which she may eat later. In Britain about fifty species of various birds have been recorded as fostering Cuckoo eggs: Hedge Sparrows and Meadow Pipits do so most frequently. In Europe there are also a large number of host species, but in certain areas one may predominate. Where this occurs, the Cuckoo lays eggs which resemble those of the host very closely. Thus in Finland they are blue, like the eggs of the Redstart, in Poland, flecked with brown like a Robin's eggs and in Hungary blotched with grey and black like those of a Great Reed Warbler. In Britain Cuckoos' eggs are not like those of either of the major hosts, but are very variable in colour.

Small birds fear the Cuckoo and try to defend their nests against it. If they detect a Cuckoo's egg among their own, they may desert the nest and build elsewhere. Because of this a Cuckoo has to lay more eggs than most other birds of her size: up to twenty-five have been recorded from one bird in a single season. The eggs are smaller than those of the host and hatch in about twelve days. The young Cuckoo, naked, blind and helpless has for the first few days of its life sensitive patches on its back, which it pushes under anything in the nest, be it egg or nestling. Bracing itself with its unfeathered wings against the side of the nest it heaves upwards and in this way it removes all competition. If it did not do so it would probably starve along with its foster brothers and sisters, for the Cuckoo is a larger bird than its hosts which find themselves hard put to it to feed the one enormous nestling. It soon outgrows the nest but remains nearby and is fed by the foster parents who may have to sit on its head to do so. It has a screaming demand call which attracts other birds with nests nearby into pushing food, intended for their own brood, into its beak.

When finally fledged, the Cuckoo feeds on various insects including hairy caterpillars, which few other birds will touch. Finally, in late Summer, it departs for its African wintering grounds, flying alone, with innate knowledge of the direction and distance it has to travel.

HOBBY —
Falco subbuteo

Found in Europe except for the far north, but only occurring in the southern counties of Britain, the Hobby, like the Peregrine, is a member of the Falcon family. It has similar long pointed wings, long tail and speedy flight. It is a smaller species, the female reaching only about 305 mm. and it prefers more open country with scattered trees or small areas of mixed woodlands. From here it hunts the birds and large insects of heath and farmland, its prey including larks and pipits, starlings and sparrows and sometimes even swifts. It hawks after dragonflies and large beetles which it holds in one foot and eats while in flight. It breeds in trees, usually in old nests of members of the crow family. The male hunts while the chicks are guarded and fed by the female. At the end of the Summer the Hobbys fly south to spend the Winter in Africa.

BUZZARD —
Buteo buteo

In Britain Buzzards are birds of the north and west

Buteo buteo

Falco subbuteo

Accipiter gentilis

Accipiter nisus

where they inhabit open rough country, rocky coasts and moorlands. In Europe they are to be found almost throughout the Continent in woodland areas as well, although they usually hunt in open places. Their broad wings and tail give them the power of a soaring flight and they may be seen circling, often high above the earth, on motionless wings, although they may draw attention to themselves with their mewing cry.

Buzzards can the ground for the small mammals which are the major part of their food. Rabbits were their staple diet until these were decimated by myxomatosis and then the numbers of Buzzards dropped with those of their prey. A further decline came because of the widespread use of insecticides but they now seem to be recovering somewhat. Unlike Falcons, Buzzards make large nests of sticks either on a rocky ledge or in a tall tree. Both parents share in the incubation and rearing of the young.

PEREGRINE —
Falco peregrinus

The Peregrine is one of the most widespread of bird species, for it occurs throughout Europe and Asia and parts of Africa, Australia and America. Many forms have been described but they are in fact all part of the same species. As with most birds of prey the female is larger than the male, and may measure up to 508 mm. long. Its habitat includes wild open country as well as forests and outside the breeding season it may sometimes be seen along the sea coast.

The Peregrine's food consists mainly of other birds, pigeons being the favourite, but many other species such as members of the crow family, gulls and sea shore birds are also taken. They are usually killed in the air, the Peregrine 'stooping' or diving on its prey with tremendous speed, and with half-closed wings, normally breaking the back of its victim with a heavy blow of the half-closed talons.

The Peregrines breed on cliffs and sometimes in trees in the old nests of other birds. The female incubates and feeds the young with what the male provides. In the past Peregrines were often trained by falconers and were quite common. More recently persecution in the name of game preservation and their unintentional destruction through insecticides has led to their becoming extremely rare over much of their range and they are now legally protected in Britain and in most European countries.

GOSHAWK —
Accipiter gentilis

The Goshawk is a rather secretive bird of forests and woodlands through most of Europe but is not a breeding species in Britain. Its rounded wings and long tail enable it to steer with absolute precision between thickly growing trees. It often perches at a vantage point and dashes out to seize some small mammal or bird, which is usually eaten on the ground, although in a place where the predator can get a good view so that it can spot enemies from a long way off.

The female Goshawk, which measures about 600 mm. long is considerably larger than her mate and appears to take a good deal of the initiative in the breeding season. Both birds take part in display flights; the female builds the nest, usually in a tall tree, often basing it on the disused nest of a Buzzard or Crow, but adding sticks and green branches to it. The eggs are incubated by the female alone, although the male catches the prey to feed her and the chicks in their early days. As they begin to grow up both parents provide the food. They have a long fledging period and do not leave the nest until about forty-three days after hatching. Even then they cannot fly and are dependent on their parents for several more days. When adult, these birds are very rapid fliers.

SPARROWHAWK —
Accipiter nisus

A bird of woodlands and coppices, the Sparrowhawk is to be found almost throughout Europe, although it is now one of the rarer birds of prey in Britain. Like the Goshawk it has rounded wings and a long tail and the female is similarly much the larger of the pair, measuring about 380 mm. Her colouring, however, is less bright than her mate's for her underparts are barred with greyish brown

Pernis apivorus

Falco peregrinus

while his are striped with chestnut bands. The method of hunting is normally to fly fast along a forest track or the edge of a wood, pouncing on any small bird which happens to be there. It sometimes flies along a hedge in more open country, popping over at intervals to catch any unsuspecting small creature and occasionally chases its quarry into the open. It has been known to kill itself when doing this by flying into the closed windows of houses. Food may be eaten on a branch or on the ground, often at a particular look-out place. The nest, built by the female, sometimes assisted by her mate, is large and bulky and usually in dense conifers. The light brownish eggs are incubated for about thirty-five days but the total time that the female is sitting is more than that since she begins before the last egg is laid. The male catches and dismembers the food for the chicks which spend up to thirty days before they are able to leave the nest.

HONEY BUZZARD —
Pernis apivorus

The Honey Buzzard, which is an occasional visitor

Scolopax rusticola

Columba oenas

Columba palumbus

tail and narrower wings, which give it an altogether slenderer appearance. Its food is much more varied than that of most birds of prey and includes some fruit and eggs as well as small birds and mammals. The main food, however, is insects, especially wasps and bumble bees. They hunt mainly on the ground, scratching their way into the nests of the insects with their feet. They catch wasps in their beaks and bite off the sting against which they are protected by the scale-like feathers on the head and the thick plumage on the rest of the body. The nest is built in a large, usually broad-leafed tree, and is decorated with green foliage. After rearing one brood they migrate in flocks to Africa for the winter months.

BLACK STORK —
Ciconia nigra

Black Storks breed through Central and Eastern Europe and in Spain, but only occur as very rare visitors in Britain, although they have been recorded

to Britain is found over most of the rest of Europe in open woodland especially where there are glades and open spaces. Up to about 510 mm. long, it soars like a Buzzard, but can be distinguished by its longer there. These large birds, which measure about 965 mm. long, are solitary inhabitants of wooded countryside where there are marshy meadows nearby in which they can hunt the small fish and amphibians which are their chief food. They also tend to occur near rivers and lakes.

The nest is usually built in a large tree in a dense part of the forest; the same site is used year by year and the old nest added to with fresh sticks and green moss which they use as a lining. During the breeding season the birds have complex greeting ceremonies and display behaviour, which often involve head and neck movements while showing the white under-tail coverts. The three or four eggs are incubated by both parents, and the young are cared for in the nest for another two months before they are able to fly. At the end of this protracted breeding season the Storks fly to their southern wintering grounds either singly or in small parties.

WOOD PIGEON —
Columba palumbus

The Wood Pigeon is found almost everywhere in Europe except for the extreme north of the Continent. It is found not only in woods and open country where it feeds and where it may be so abundant as to be a pest of agricultural land, but also in parks, gardens and towns. Here it loses the waryness which enables it to survive in the country and may be seen with flocks of feral town pigeons showing no fear of man. It feeds on a wide range of mainly plant food, often plundering and stripping fields of green vegetables or freshly sown seed, as well as taking grain shed at harvest time or even small potatoes. In the woods its food includes acorns and other wild seeds and a certain amount of insect food and snails. Its nest is a fragile platform of sticks, in which two white eggs are laid. The squabs, as with all young pigeons, are fed at first on 'pigeon's milk', a crop secretion produced by the parents. As soon as they are reared other broods are produced through the summer months. In some parts of their range Wood Pigeons are migratory and the winter population of these birds in Britain is swollen by migrants from the colder parts of the Continent.

Ciconia nigra

Streptopelia turtur

STOCK DOVE —
Columba oenas

This bird often joins flocks of Wood Pigeons in the wintertime, but is smaller (about 330 mm. long) and darker in colour, lacking any white on its neck or wings. Its flight tends to be more rapid, and when feeding it normally eats the seeds of various weeds especially fat hen and knot grass in preference to the seeds of cereal crops. Its habitat is usually open woodland or parkland from southern England to Central Asia. It has a gliding display flight which may be seen during the breeding season which, like that of other pigeons, is lengthy, for it produces several small broods. The two eggs are laid in a hole in an old tree, perhaps one originally used by a Woodpecker.

TURTLE DOVE —
Streptopelia turtur

Bushy country with small woods and overgrown hedges is the most likely place to find the Turtle

Tetrastes bonasia

Dove, an elegant bird about 330 mm. long which migrates to Northern Europe from the Mediterranean region in April. The birds ase usually seen in pairs or small flocks. It occurs in southern Britain but is absent from Scotland and Ireland, perhaps because of a lack there of the fumitory, a plant which forms its favourite food although it will eat other small weed seeds and occasionally invertebrates. Abroad it ranges across Europe south of Scandinavia and may be found as far east as Central Asia. The nest is a slapdash platform built by both birds, usually fairly low in dense vegetation. Both incubate the eggs and rear the young which are quickly followed by a second brood.

WOODCOCK —
Scolopax rusticola

Although it is about 355 mm. long, the beautiful dead leaf colours of the Woodcock camouflage it completely in the forested regions from Britain and France across Europe through much of Northern Asia. During the daytime this secretive bird remains in thick cover but emerges to feed in the evening on worms and other small invertebrates. It pulls these from the ground with its long beak in the manner of its relatives the Waders of the tundra and estuaries. Its normal flight is rapid, but during the breeding season, between March and July, Woodcocks tour the boundaries of their territories at dawn and dusk in the 'roding' flight. In this, they fly slowly, calling with a high-pitched two syllable whistle interspersed with a grunting croak.

The Woodcock's nest is made of leaves and moss in a shallow scrape at the base of a tree. The female incubates the four cream and brown coloured eggs for about twenty-two days and cares for the chicks after they hatch. If danger threatens she may airlift them for a short distance holding them between her thighs as she flies. Over part of their range Woodcocks are migratory, and those birds in Britain are augmented in wintertime by flocks which come in from the Continent, often spending the first day in marshy fields before flying off to more suitable places to live through the cold weather.

BLACK GROUSE —
Lyrurus tetrix

The male Black Grouse, which is about 530 mm. long, is known as the Black Cock while his smaller and duller coloured mate is called the Grey Hen. In Britain it is found in coniferous woodlands in upland areas of the north and west; in Europe it occurs widely from eastern France across the Continent to Central Asia but not in Italy or the more southerly areas. In Winter it may be found in open birch woods. In early Spring the Black Cocks assemble at their display grounds or leks which are on open ground close to the woods. Here soon after day break they posture and spar before the Grey Hens which sometimes also fight among themselves. Mating may take place at the lek or away from it, but the species is polygamous and the cock takes no part in rearing the brood. The chicks are able to run about shortly after hatching, and can fly fairly soon after this. Black Cocks feed chiefly on the buds of trees, especially birch and pine which earns them the enmity of foresters, although they also eat seeds and berries, some insects and other invertebrates.

CAPERCAILLIE —
Tetrao urogallus

The Capercaillie is found in upland areas from Scandinavia into parts of Eastern and Southern Europe and in the Pyranees. It used to be native to Scotland becoming extinct there at the end of the eighteenth century and so was reintroduced into Perthshire in the

Phasianus colchicus

Lyrurus tetrix

Tetrao urogallus

1830s. From that time on it has flourished and is now fairly common over much of the highlands wherever there are well grown coniferous woodlands. It is a large bird, the male measuring more than 860 mm. long while his mate is some 250 mm. smaller. During the breeding season the males become highly aggressive and display on branches and on the ground with an extraordinary hiccuping song which ends with a noise like a cork being drawn from a bottle; this is then followed by a spell of hissing noises. The nest is made on the ground at the foot of a large tree and the eggs and young are tended by the female which is well protected by her camouflaged colouring. The male, which is polygamous, takes no part in rearing the family. Capercaillie feed mainly on the shoots of conifers which makes them very unpopular with foresters, for they defoliate the trees to some extent and if they remove the leading shoots the trees grow crooked and become worthless as timber.

HAZEL HEN —
Tetrastes bonasia

The Hazel Hen is a woodland species found mainly in forests in upland areas. It prefers places where there is dense ground cover of bilberry or some such shrub where it can hide, although it perches in trees quite readily. About 335 mm. long, it is now less common

63

than formerly, partly because of destruction of its habitat and partly because in common with most of its relatives it makes excellent eating and is widely hunted, although it is still numerous in Scandinavia. The eggs are laid near to the base of a tree and the young which hatch after an incubation period of twenty-five days are soon independent, although they remain with their mother for warmth. They feed on shoots, seeds, berries and small invertebrates including spiders, insects and snails.

PHEASANT —
Phasianus colchicus

Pheasants are found in woodlands throughout Britain and most of Europe, except for the hottest and the coldest and most mountainous areas. The male measures up to 835 mm., over half of which is the length of his tail. He is brightly coloured but his mate wears more sombre feathers and has a shorter tail; her length is a third less than his, mainly because of this discrepancy.

The Pheasant originated in the Middle East. It is reputed to have been brought to Britain by the Romans but there is no proof of this although it was certainly here before the Norman Conquest. Pheasants fly strongly but not for long distances which makes them favourite game birds.

Repeated reintroduction of different interbreeding stocks has now made the Pheasant among the most variable of species; among the many colour forms the Ring-necked Pheasant is the variety most often met with. The males are polygamous and the hens incubate and care for the chicks which are active as soon as they are hatched. Pheasants eat grain, leaves and invertebrates, but although they feed on the ground, often in the open, they return to the trees in which they roost at night.

MAMMALS

NOCTULE —
Nyctalus noctula

The Noctule, which is found almost throughout Europe except for northern Scandinavia, northern Scotland, Ireland and north west Spain, is the largest British bat, with a body length of up to 82 mm. and a wingspan of up to 390 mm. It normally rests in holes in trees from which it emerges to feed early in the evening and occasionally even during the daytime.

The Noctule may be recognised by its long slender wings and red-brown colour. It flies fast, up to 100 metres above ground level, where it catches beetles, moths and various other large insects. Noctules are highly social and up to 300 have been found in one noisy colony. In summertime the colonies may shift from one tree to another but during the Winter large groups may migrate from Northern to Southern Europe.

COMMON SHREW —
Sorex araneus

Although absent from Ireland and most of Spain, the Common Shrew is found almost everywhere in Europe in a wide variety of habitats, including rough pasture and meadows as well as forests. Occasionally it even enters houses. It may be recognised as a shrew by its long snout and small ears and eyes and as a Common Shrew by the body length of up to 85 mm. and tail length up to 56 mm. with a definite stripe of lighter brown between the dark brown colour of its back and the ash-grey of its belly. Shrews are active through the day and night, hunting the small invertebrates which they must eat in quantities exceeding their own body weight every twenty-four hours if they are to survive. Their shrill voices may often be heard as they rustle through the undergrowth. The females normally produce two litters during the Summer, but die after a short life which never exceeds just over one year.

RABBIT —
Oryctolagus cuniculus

Rabbits originated in Spain from where they have spread, partly at least through human agency, to much of Europe as well as other parts of the world such as Australia. The Normans are credited with

64

bringing them to Britain in the twelfth century. They are absent from most of Scandinavia, Italy, the Balkans and Eastern Europe for they cannot tolerate mountain environments and are chiefly animals of the lowlands. Here, however, any place where it can dig a burrow is home to a rabbit, and woodlands, especially coniferous woods, heathlands and open country are all likely to house the species. Rabbits are social animals and the burrows which may form a complex series of tunnels are grouped together into warrens. Activity may be seen here at any time of the day but is greatest at dawn and dusk, when the animals do most of their feeding. For this they do not normally move very far from their homes, but graze grass and a wide variety of herbs, nibbling them very close to the ground. In this way they may prevent the regeneration of trees in an open area and in a hard Winter they may kill large trees by nibbling the bark. Female rabbits mostly give birth to their young in the Spring and early Summer. During this time several litters are produced of naked, blind, helpless young, mostly born in a special den excavated by the mother a short way from the main colony and lined with grass and fur pulled from her underside. The young are able to leave the nest at about twenty-one days, and are sexually mature at about three months, before they are fully grown. At full maturity a rabbit will measure not more than 450 mm. and weigh about two kilogrammes at most. Until 1953 the rabbit population of Europe was very high, but in that year myxomatosis, a virus disease, was introduced from South America. The rabbits had no defence against this infection and the first wave of the disease caused a very high level of mortality. A few escaped, and populations with some resistance are being built up in many areas, although myxomatosis is now endemic and likely to affect rabbit survival for many years yet.

RED SQUIRREL —
Sciurus vulgaris

Through most of Britain the common squirrel is the American Grey Squirrel, a species which is unknown in the wild in mainland Europe. There the Red Squirrel, which still survives in western and northern Britain and part of East Anglia, is the only species,

other than the Flying Squirrel of the forests of the far north. The Red Squirrel may be recognised apart from its bright rusty colour, (although a black form of the species is common in the mountainous areas of Southern Europe) by its small size, for its head and body measure only about 220 mm. and the tail adds another 180 mm. to this. At all times the ears look pointed for they carry long tufts of hair. Red Squir-

Nyctalus noctula

Oryctolagus cuniculus

Sciurus vulgaris

Glis glis

Apodemus flavicollis

Muscardinus avellanarius

rels are most at home in coniferous forests but they exist wherever there are trees, up to over 2 000 metres in the Alps. They make nests or dreys of twigs in the branches although in Winter they may make use of a hollow tree for shelter. Each female normally produces at least two litters of young in the summer months. Squirrels feed on a wide variety of things; mainly seeds, fruits and fungi, but also insects and birds' eggs and their young and in the wintertime they often do a great deal of damage by nibbling the tender growing shoots of the trees. Their enemies include birds of prey, foxes, martens and man who kills many of them each year, mainly for their pelts.

EDIBLE DORMOUSE —
Glis glis

The Edible Dormouse looks at first sight like a small Grey Squirrel, but the large eyes and small ears should serve to distinguish it. It is the largest of the Dormice. It is known from Central and Eastern Europe but it was introduced into the Tring area of Britain late during the nineteenth century and has survived there without much increase until recently, when there have been signs of large numbers of them invading the woodlands. It lives semi-socially, preferring deciduous woodlands, gardens and orchards, feeding on seeds, shoots and fruit, especially apples. It sometimes enters houses and apple stores where it may be heard rather than seen, for it is strictly nocturnal. In woodlands it is very agile and climbs and jumps about the trees. The young are born in early Summer in a nest made in a hollow tree. In the time of abundant food in the Autumn, the Edible Dormice

become very fat and well before the onset of Winter they creep into holes in the ground, warmly lined with moss and grass, and go into the deep sleep of hibernation until Spring is well advanced.

DORMOUSE —
Muscardinus avellanarius

The beautiful orange-brown Dormouse used to be fairly common in British woodlands, but is now becoming increasingly rare there. It is present almost exclusively in the West Country and Wales although it is found across Central and Eastern Europe especially where there is plenty of low scrub for cover. In spite of its plump appearance it is very agile and can run and jump through the trees at night which is when it is normally active. Its summer nest is usually above ground, often using the papery bark of honeysuckle as a base. The winter nest, in which it hibernates, is normally in a protected position on the ground. The young are born in the early Summer but hibernate just as their parents do from early Autumn until Spring.

BANK VOLE —
Clethrionomys glareolus

Voles are different from mice in their blunt faces, small eyes and ears and short rather hairy tails. The Bank Vole which lives throughout almost all of Europe except for northern Scandinavia, Spain, most of Italy and Ireland, is variable in measurement according to locality, but the maximum head and body length is about 120 mm. with a tail length of about 72 mm. It lives in warm dry places where there is plenty of cover in the edges of deciduous woods, hedgerows and gardens, making shallow burrows and a warm ball-shaped nest of grass and moss. Its food consists of a variety of seeds, fruit and leaves, and in times of plenty it may make a larder of stored food in its home tunnels. This it eats during the wintertime, for unlike the Dormouse, it does not hibernate. The female produces a series of litters of blind, helpless young through the summer months but is unlikely to live for more than one breeding season.

YELLOW-NECKED MOUSE —
Apodemus flavicollis

The Yellow-necked Mouse is found in dense woodlands and scrub in Central and Eastern Europe, the
Balkans, parts of Scandinavia and southern Britain.
It is strictly nocturnal, but if seen may be recognised
by its large size, (for its head and body length add
up to 130 mm. with a tail length exceeding this) and
the collar of yellow fur round the neck which shows
clearly against the pure white of the underparts.
It is an excellent climber and jumper and has been
recorded as living in the tops of tall trees.
The Yellow-necked Mouse makes its nest in a
variety of places, including bird nest boxes. In
Britain it is more likely to enter houses than its
cousin the Woodmouse, and may even raid apple
or bulb stores. In the wild its food is very varied,
including many sorts of seeds and fruits and some
insects, especially in the early Summer. In the Autumn
the Yellow-necked Mouse makes stores of food for it
does not hibernate through the winter months. The
females produce several litters of young during the
Summer, but their numbers are normally held in
check by their many enemies which include foxes,
stoats, hawks and owls.

WILD CAT —
Felis silvestris

The Wild Cat, which is found in more or less wooded
parts of the mountainous areas of Southern and
Western Europe and northern Scotland, looks like
a large, rather short-tailed tabby. It is not, however,
the ancestor of the domestic cats, although it is closely
related and will interbreed with them. It is a solitary
animal for most of the year, and the male does not
help with rearing the young which are born in a
secure den in an inaccessible place among the rocks
or a tangle of fallen trees. The Wild Cat does not
dig its own shelter, but may use a fox's burrow. The
kittens remain with their mother for several months
during which time she cares for and defends them
against all enemies. The cats are mainly active at
night, hunting mice, small birds, lizards and even
insects and in exceptional cases may attack larger

Felis silvestris

Lynx lynx

Canis lupus

animals such as hares. Wild Cats, like most other
carnivores, used to be common over a far larger
area than they now inhabit, but have been driven to
the more remote places by human persecution. In
some places they are now increasing in numbers and
in Scotland are now being seen in areas from which
they have been absent for many years.

67

LYNX —
Lynx lynx

The Lynx, which is the largest member of the cat family surviving in Europe, is widespread but nowhere common, found in the remoter mountain areas of Northern, Eastern and Central Europe. It is a circumpolar species and also occurs across Asia and North America. The smaller, more heavily spotted Spanish Lynx is one of the world's rarest mammals, surviving only in parts of Spain and South East Europe. Lynxes are solitary animals, the male leaving the female after a brief courtship, although the kittens which are born in May remain with her for many months. The den in which they are born is usually well camouflaged and inaccessible for Lynxes have learnt that the presence of man usually bodes ill for them. Their excellent hearing and eyesight ensure that they keep well out of the way of human beings. Lynxes normally hunt at night, feeding on larger prey than does the Wild Cat, up to the size of chickens and young deer, which makes them an enemy of farmers wherever they live.

WOLF —
Canis lupus

The wolf is among the most feared and hated of predaceous animals, with the result that it has been exterminated from many of its former haunts throughout Europe. It now occurs abundantly only in North East Europe; although it survives in Central Europe, Germany and Italy it is very rare. It is a circumpolar species and is found also in parts of Asia and North America but here again it has been reduced to very low numbers. Where it is found it occupies woods and steppeland and tundra and mountains up to a level of 2,500 metres.

Intelligent and social, Wolves live in family groups and hunt their prey by running it down. Their natural food includes large mammals, such as deer as well as many smaller creatures, including insects and molluscs. They will also eat a certain amount of vegetable material. Wolves are not wasteful or wanton killers and will normally finish up completely the carcase of any victim. They often

Martes martes

Vulpes vulpes

Meles meles

travel very long distances in their search for food and the territory of a pack, which is marked out with urine, may cover several hundreds of square kilometres.

This habit of territory marking is one which they have passed on to their descendants, the domestic dogs. Wolf cubs are born in April and May in a burrow which may be dug under logs, tree roots or rocks. At first they are blind and helpless and their mother remains with them all the time, while other members of the pack bring her food. Later they learn the skills of hunting by working with their parents with whom they remain for at least two years.

FOX —
Vulpes vulpes

Foxes are to be found throughout most of Britain and Europe in a wide range of environments including the suburbs of many major towns, where they make a living largely by scavenging garbage. Elsewhere they feed on small mammals and birds, insects, worms, carrion and a good deal of vegetable material. The Fox's earth may be an enlarged rabbit hole or a dry place under the roots of a fallen tree. It is possible to tell whether it is in use for the Fox is one of the few animals which fouls its home and produces an unmistakeable, powerful smell which lingers for several hours. The cubs are born in Spring and reared by both parents, who teach them the arts of hunting by bringing to the den a disabled animal such as a rabbit or moorhen on which the young can practise their skill.

PINE MARTEN —
Martes martes

An animal of coniferous and mixed forests common over most of Europe, but now becoming increasingly rare. In Britain it occurs only in the remoter mountainous areas of the north. Although it is very shy and unlikely to be seen, if it should be glimpsed it may be recognised by the bib of yellow coloured fur beneath the chin. In the tree tops it is astonishingly agile, able to leap several metres and to cling to small branches with its large flexible paws. It hunts at night and feeds on squirrels, small birds, insects and fruits, loving to

Ursus arctos

eat sweet things. The young are born in Spring in a den on the ground and do not climb until they are well grown.

BADGER —
Meles meles

The Badger, which lives in Britain and Europe except for the most northerly parts, is a shy and secretive animal, strictly nocturnal, so that even where it is fairly common it is rarely seen. Evidence of its presence may be seen in the large dens, called sets, which they dig in sheltered areas. These may often be very extensive, stretching for as much as 100 m. along a hill slope with a dozen or more entrances and exits. Several animals may occupy one set but they do not appear to live socially except for a brief courtship period. Badgers are very clean animals and keep their sleeping quarters lined with dry grass or bracken which they replenish at frequent intervals. The paths leading to and from their sets have probably been used by many generations of Badgers for they are animals with a very fixed routine when they go out foraging for the plants and animals on which they feed. Although Badgers are powerful creatures they do not normally attack anything beyond small rodent or rabbit size, feeding largely on invertebrates and roots, shoots and bulbs and having a particular liking for

Sus scrofa

anything sweet. The female gives birth to up to five cubs in February but these do not emerge from the set until late March. The male takes no part in their upbringing.

BROWN BEAR —
Ursus arctos

Brown Bears are circumpolar animals, occurring in Europe, Asia and North America. In so large an area many forms have been described varying in size and other details; the largest are the Kodiak Bears of Northern America. Brown Bears used to be found in Britain, but were exterminated there in the twelfth century. Elsewhere in Europe the species is now confined to large forested areas mainly in the north and south east of the Continent. Its den is usually excavated under uprooted trees or among rocks. Here the female gives birth during the Winter to two to five cubs. These are blind, naked and helpless and weigh only about 100 grammes compared to their mother's weight of over 100 kilogrammes. The female does not stir from the lair but remains with her young, suckling them and keeping them warm for several weeks and although as Winter retreats she leaves them to feed, they do not venture out until they are about three or four months old. Bears are omnivorous, eating fruits, roots, insects, small mammals, birds and their eggs, carrion, fish and if they can possibly get it, honey. They very rarely attack large animals.

WILD BOAR —
Sus scrofa

Like most of the larger animals of Europe, the Wild Boar is now much rarer than in former centuries, when it lived, among other places, in Britain. It is

widespread through forested country from Spain to Eastern Europe, though the highest mountain regions seem to be inimical to it. In spite of the fact that they are hunted as game animals, in recent years they have increased slightly in numbers. The males are generally solitary; the females live in small groups with their young, which are born in a secure den warmly lined with grass and moss. At birth the piglets are striped but they lose this camouflage as they grow up. Wild boars are omnivorous, feeding on berries, roots, grass and sometimes raiding fields of root crops. They also eat any invertebrates, birds' eggs, and carrion they can find and will kill and devour snakes, even the poisonous species.

FALLOW DEER —
Cervus dama

Originally an inhabitant of the Mediterranean region, the Fallow Deer has spread widely through much of Europe. In the northern parts of its range this increase is due to man: the Romans are credited with bringing the Fallow Deer to Britain, for example. As with many animals controlled by man, there has been selection for various colours and a herd of these deer may include individuals which are very dark or almost white as well as the reddish-brown colours spotted with white, which is the normal shade. Fallow Deer are found mainly in open woodlands with grassy clearings, for they graze as well as browse on the trees and shrubs and they may invade fields to feed on root crops. As they are highly social several deer will be involved in such a venture and so considerable damage may be done in this way. The rutting period is in October and November. The males make a grunting call and as well as fighting spread their scent with their palmate antlers. The fawns, usually born singly, are dropped in May or June.

RED DEER —
Cervus elaphus

Red Deer are to be found in large areas of woodland through much of Europe. Although they can survive on open ground, as they do to a large extent in Scotland, under such conditions they are smaller and

70

lighter in weight than the forest deer. This species occurs through much of Asia and North America where it is called the Elk or Wapiti. It is a mainly nocturnal herd-living animal, the mature stags normally living apart from the hinds and juveniles which have a more stable social structure. During the rut in October the stags roar and fight, using their heavy branched antlers and marking their territories with the secretions from facial glands. The hinds drop their calves in June. These are spotted and lie hidden in the undergrowth while their mothers feed. They are suckled for the greater part of their first year, although they start to eat the leaves, grass, berries and fungi of their habitat when they are only a few weeks old. Although the young may be the prey of many predators, the adults have no real enemies other than man, who has preserved and poached them throughout the centuries.

ROE DEER —
Capreolus capreolus

Although some very small forms of deer have been imported into Europe, the Roe is the smallest of the native species, standing only about 700 mm. at the shoulder. It is widespread in deciduous and Mediterranean woodlands and also occurs to some extent in the coniferous forests of the north. It is a less social animal than the other European deer, normally being seen singly or in small groups. The rut is in July and August when the males mark out their territories by rubbing the bark of saplings and shrubs with their antlers as they spread the scent from facial glands. Well-marked deer rings formed in the course of chases round a particular tree are another indication of the presence of these shy and largely nocturnal animals. The doe produces her young — usually twins — in May and June and these remain with her until the next year. Roe feed on a great variety of plants, and often damage young plantations by eating the growing shoots of the saplings. In Britain where they have increased in recent years they may enter rural gardens where they find roses and bean plants particularly to their liking. Where they occur, wolves are their main enemies apart from man, although many of the kids are taken by foxes.

Cervus dama

Cervus elaphus

Capreolus capreolus

LAKES,
PONDS
AND
POOLS

INTRODUCTION

Much of Europe has a high rainfall and ponds and lakes which trap a proportion of this water are a common feature of the countryside. The exceptions are in the dryer parts of the south, and areas where porous limestone rocks at the surface allow water to soak through to form underground streams, often of considerable extent.

Lakes and ponds may be formed in many ways. They may be man-made to beautify the landscape, may be present as a result of quarrying, may have been dug for the purpose of breeding fish, or giving water to farm stock, or as reservoirs. Natural bodies of standing water may form where the water table is high making shallow ponds in low lying places and deep lakes in upland areas, especially where these have been heavily glaciated. The differences between ponds and lakes are essentially those of depth. In a pond the water acts as a more or less uniform body, heating or cooling according to the season. In a lake the surface water warms during the summer months and cools during winter, but below this is a body of water in which the temperature varies little; it is relatively warm in winter and cold in summer. The junction between the upper water of variable temperature and the lower, stable water is called the *thermocline*. The deep, cold, dark water below the thermocline contains relatively little life, although the upper waters may carry numerous microscopic plants which are the basis of food chains culminating in fishes, birds and aquatic mammals. In a pond light may penetrate through all the water and life of microscopic and larger size occurs throughout; rooted plants may grow from any part of its bed. Ponds are often very rich in nutrients since the streams that feed them drain rich lowlands. Lakes frequently contain rather few nutrient salts since their tributary streams run over old, hard, insoluble rocks.

Ponds and lakes are both transient features of the landscape, for the streams which bring them water also carry silt, which is deposited in the water. Silting is aided by the growth of plants which trap mud in their roots and stems, encroaching further into the water as it becomes shallower until finally it disappears altogether. Artificial lakes may need to be dredged to maintain their size; small ponds may disappear within the range of human memory, and even large lakes are not immune. In Britain the region of flat land known as the Vale of Pickering in Yorkshire was in prehistoric times a lake, now completely silted over, and in the English Lake District Buttermere and Crummock Water, now two separate lakes, were created by the encroachment of river deltas into a single large lake which once filled the whole valley. Many other such examples may be found through Northern and mountainous Europe. The destruction of lakes may not always be natural. Farmers wishing to use every corner of their fields may fill in ponds: old quarries containing water may be filled so that the land my be reclaimed.

Furthermore the life of ponds and lakes may be totally destroyed by the uses to which they are put. In small ponds, the process of eutrophication (over enrichment) may take place with the run off from the land of fertilising chemicals used by famers. This leads to an overgrowth of minute, short-lived plants, which as they die, remove the oxygen from the water in the course of their decay. Once this has happened the animals and other plants die, leaving the water quite lifeless. In larger bodies of water eutrophication may also occur but pollution from sewage and industrial effluents may accelerate the process of destruction of the lake as a living entity. Many of the great lakes of South Central Europe are now very seriously polluted and contain little life compared with former days. In spite of this, there are still many stretches of water in which the animals described in this text may be observed.

Animal life is always dependent on plants. Ponds and lakes have their own flora which differs from that of other ecological regions. At the approach to a pond plants which can stand a certain degree of waterlogging can be found. On the edge of the water rooted plants such as reedmace and yellow flags, may live with their lower growth entirely submerged. In deeper water some rooted plants still occur such as the beautiful yellow or white water lilies, or the fragile-leafed pondweeds. Finally there are some floating plants such as the small water ferns, duckweeds or frogbits which have no roots and obtain all of their nutrients directly from the water. Beyond this microscopic algae may be abundant, in some cases forming a green scum over the surface of the water. Animal life is generally richest where the plant growth is most abundant and varied, offering food, hiding and living spaces to a wide range of species inter-related in their ways of life. In Northern and Central Europe, most ponds and lakes freeze in Winter and active life in the water may thus be curtailed for several months.

WORMS

RIVER WORM —
Tubifex sp.

The fine mud in shallow ponds and slow flowing streams may sometimes have a velvety-red appearance. Vibrations from the bank or a sudden movement in the water will cause this to disappear, leaving only tiny pits visible. These are the homes of the River Worms, which burrow, head downwards, into the silt, feeding on the organic matter which it contains. They are sometimes found in areas which are deficient in oxygen so few other creatures can survice. The reason for this is that the worms contain haemoglobin which has a great affinity for oxygen. This means that they can capture whatever small amounts of it may be in the water. Haemoglobin which is red, gives the worms their bright colour, as it does to the blood of vertebrates where it is also present. Like their close relatives the Earthworms, *Tubifex* are hermaphrodite and exceedingly prolific. They are collected by aquarists for use as fish food.

FISH LEECH —
Piscicola geometra

Leeches, which are able to change shape, becoming enormously elongated as they move, have at first sight little in common with any other animals but their segmented bodies show their relationship with Earthworms and *Tubifex*. They are relatively inactive but can hold on to their prey with two suckers, a large one at the tail end and a smaller one near the head. The Fish Leech which reaches a length of about 400 mm., sucks blood from fishes which in a heavy infestation may be killed by the parasites.

HORSE LEECH —
Haemopis sanguisuga

The Horse Leech, which may be green or brown spotted with black, is common throughout Europe in ponds, ditches and sluggish streams. It measures about 125 mm. long but in spite of its size and name it is quite harmless to man or any other mammals.

It does not suck blood but attacks and swallows snails and other invertebrates or tiny frogs and fishes. It is sometimes found out of the water for it deposits its parchment-like egg cocoons in damp earth.

MOLLUSCS

RIVER SNAIL —
Viviparus viviparus

The River Snail occurs mainly in weedy, slow flowing rivers and ponds in Britain as far north as Yorkshire and across much of Central Europe and into Asia. Feeding on small water plants and organic material from the mud, it may grow to a shell height of 50 mm. On the upper surface of the foot towards the rear end of the animal, the operculum may be seen. This is a horny disc which is used to close the aperture when the snail retreats into the safety of its shell. The eyes are at the base of the single pair of tentacles which cannot be retracted in the way of many snails. The sexes are separate and may be distinguished by the shape of the tentacles. In the females these are both long and tapering, while in the male the right tentacle is short and thick and is used as a mating organ. River Snails do not lay eggs but produce large numbers of small young in which the shells are already formed.

LISTER'S RIVER SNAIL —
Viviparus fasciatus

This species is closely related to the River Snail, which it resembles in many respects. It is, however, slightly

Viviparus viviparus   *Limnea stagnalis*

smaller, growing to a shell height of about 30 mm. and has a wide distribution over much of Europe although in Britain it only occurs in the south. The shell colour is usually dark grey, spotted with black, but many colour varieties have been recorded.

GREAT POND SNAIL —
Limnea stagnalis

Great Pond Snails are found in lowland ponds throughout Britain and Europe into Western Asia and North Africa. A large species, its fragile shell grows to a height of 60 mm. and a width of 34 mm. Although they are always found in water they do not have gills but breathe by means of a lung which they must recharge at the surface. They often crawl on the underside of the surface film breaking through in order to take a breath. At this time the circular breathing hole is opened on the right side of the body behind the snail's head. Pond Snails have the surprising habit of changing their level in the water, sinking rapidly or suddenly floating up to the surface again. They do this by compressing the air in their lungs and re-expanding it at will.

The food of Pond Snails is varied, including large water plants as well as micro-organisms and algae. It also eats some animal material including living small fish or newts. The eggs are laid in gelatinous masses, attached to water plants or stones. When the young snails hatch they seem at first to have a very poor ability to hang on to anything and are easily dislodged and swept downstream, even in very placid waters. This is a method of dispersal for an otherwise very sedentary species.

When adverse conditions, such as drought, cause their ponds to dry up Great Pond Snails can protect themselves by producing a layer of mucus across the shell opening, which hardens and prevents the animal from dessicating. This is quite different from the horny operculum of the River Snail which is present throughout the life of the animal.

DWARF POND SNAIL —
Limnea truncatula

This small relative of the previous species rarely grows to a size of more than 10 mm. high and 6 mm. wide. It is generally found in shallow water or even on marsh plants at the very edge of the pond which is where its small clumps of eggs are laid. It may become very abundant for development is rapid and several generations can be produced in a single Summer. In some areas it is of great agricultural importance, for it is the intermediate host of the Liver Fluke which parasitises sheep and cattle causing loss of condition and occasionally death to the infected stock. Sometimes humans may become infected, via the animals parasitised, by the Liver Fluke, usually with severe results.

GREAT RAM'S HORN —
Planorbis corneus

Easily recognised by the flat spiral of their shells, a number of species of Ram's Horn Snail are to be found throughout Europe. The largest is the Great Ram's Horn which measures up to 35 mm. across. This is widely but rather locally distributed across Britain and much of Europe, extending to Siberia in the east. It is usually found in weedy places, where there is plenty of food.

In times of hardship it can protect itself with a layer of hardened mucus in the same way as the Pond Snail. Perhaps because of this, it often survives for three years, a longer life than that of most other water snails. A feature which separates them sharply from all their close relatives is the presence of haemoglobin in the blood, which is red, like that of vertebrates.

AMBER SNAIL —
Succinea putris

This snail, which is found mainly in damp meadows throughout Britain and Europe, may measure up to 27 mm. in height in some eastern parts of its range, although it is generally smaller. Like many snails it is the intermediate host of a parasitic fluke which in this case infects birds. These attack the snail's antennae which have become brightly coloured by the presence of the flukes. The snail does not appear to be disconcerted by their loss, but merely grows another

80

pair. They may live for three years, laying their eggs in damp meadows.

SWAN MUSSEL —
Anodonta cygnea

There are fewer species of bivalves than of snail-type molluscs to be found in fresh water. However, those that are present may grow to a large size. The Swan Mussel, for example, which is widespread in slow flowing waters and large ponds over lowland Europe is recorded as growing up to 250 mm. long although half this length is a more usual size. These animals lie half buried in the mud, filtering tiny food particles from the water around them. During the Summer the females produce very large numbers of tiny eggs. These are not spawned into the water, but are retained in brood pouches on the gills where they are fertilised by sperms drawn into the body along with food. The larvae remain in the shelter of their parent's shell for some months but when they leave they are parasitic on fishes for a short time. Here they complete their development and finally take up a sedentary life at the bottom on the water.

LAKE ORB MUSSEL —
Sphaerium lacustre

This little mussel, which rarely grows beyond a length of 10 mm. is nearly circular in shape and so may often be mistaken for a pebble. It is found widely in lowland ponds and slow-flowing rivers through much of Europe and Asia. The Orb Mussels are hermaphrodite, each individual producing young which are fully formed when they leave their parent's body so they do not have to undergo a larval or parasitic stage of life.

SPIDERS

WATER SPIDER —
Argyroneta aquatica

Although many spiders are to be found in damp places near the edge of ponds and streams, only one is an actual inhabitant of the water. It is found widely through Europe in weedy pools in areas which are unfrozen for most of the Winter. Although striped, under water it looks silvery for its body is clothed in short fine hairs and these trap tiny air bubbles when it comes up to the surface. It is unusual among spiders, for the male is slightly larger than his mate, which measures about 8—15 mm. long. All spiders must have air to breathe and the Water Spider ensures its underwater supply by building a closely spun raft-like web between the stems of water plants. It then goes to the surface and brings down bubbles of air. These are trapped underneath the web, which bulges upwards to form a diving bell in which the spider spends much of its time. The female puts her egg cocoon at the top of the bell which is the nursery of the tiny spiders. Water Spiders feed mainly on aquatic insects and their larvae which they usually take back to their bell to eat.

CRUSTACEANS

WATER FLEA —
Daphnia pulex

In ponds and the upper water of lakes microscopic

Succinea putris

Planorbis corneus

Sphaerium lacustre

Anodonta cygnea

plant life, which is the food for many small animals, is abundant through the summer months.

One of the most important of these is the Water Flea which may at certain times be incredibly numerous. These creatures which are at most only about 2 mm. long are not, as their name suggests, insects, but crustaceans, cousins of the crayfishes and the shrimps. Their bodies are enclosed in a transparent carapace through which the movements of the limbs and the working of the internal organs may be seen. More easily visible is the single large eye.

Water Fleas swim by beating their branched antennae and it is this jerking action which has given them their name. They are often pink in colour for their blood may contain haemoglobin which enables them to use any oxygen in the water very efficiently. Almost all Water Fleas are females, capable of reproducing without mating and giving rise to further generations of females like themselves. Under certain circumstances, however, usually those which occur as winter approaches or the pond is about to dry up, a generation of males is born. These mate with the females, who then produce eggs in well protected cases, which are able to withstand both freezing and drought. These then hatch into more females very speedily as soon as conditions are suitable.

WATER HOGLOUSE —
Asellus aquaticus

Another group of crustaceans which are widespread in Europe and often very abundant in weedy pools and slow-flowing streams are the Water Hoglice, represented here by *Asellus aquaticus*. There are a number of species of these scavengers, all very similar in appearance and reaching about 25 mm. as a maximum length.

Water Hoglice's bodies are slightly flattened from above, and they have a large number of pairs of legs, attached to the segments of the thorax and abdomen, which enable them to move remarkably fast when they wish. In Winter *Asellus aquaticus* may often be found in pairs, the male holding the female beneath his body. In Spring the females carry their eggs in a mass below the front end of their bodies. The young resemble their parents on hatching and do not have to pass through any separate larval stages, but they remain with their mother for a short time.

INSECTS

WATER SPRINGTAIL —
Podura aquatica

Sooty looking patches at the edge of still water may, if disturbed, sometimes break up into thousands of tiny leaping creatures never more than 1 mm. in length. These are Springtails, which are among the most primitive of insects.

A Springtail never has wings but is able to jump many times its own length by using a forked 'vaulting pole' which is normally tucked up under its body, and held by suckers between the back legs. When this is released the animal springs several centimetres into the air, a defence mechanism which must protect them against many predators. Once any danger is past the group of Springtails will reassemble. When not alarmed, Springtails move slowly over the surface film of the water, sometimes crawling down plant stems below the surface but never getting wet as their bodies carry a protective film of air. Little is known about Springtails' way of life but they probably feed on decaying plant matter on the surface of the water.

MAYFLY —
Ephemera vulgata

Many species of Mayfly which may be recognised by the way they hold their four gauzy wings erect over the back, are known from all over Europe. Some occur in fast flowing waters while others, such as the one illustrated, are characteristic of ponds and slow rivers. All start their lives in water and may spend up to two years in the nymphal stage, when they may be identified by the three long tail bristles which they carry. They are mainly plant eaters but may sometimes scavenge or feed on other small animals.

After their prolonged aquatic life the Mayflies survive for only a short time once they have left the water. According to the species this occurs during the early or midsummer months when huge numbers of the insects moult their last aquatic skin and leave the water to fly away for a few hours. At this time they are still not fully mature and are a dull colour. They are referred to by fisherman as Duns but within a short time they have their final moult into the bright adult colours. They then return to the water, where the males congregate in great swarms, rising and sinking over the water in a dancing flight. Any female which appears quickly finds a mate and lays her eggs in the water. Once this is accomplished the adult insects die, after an aerial life of less than a day. At all stages of their lives Mayflies are important food for fishes and the artificial flies made by fishermen in many cases imitate their salient features.

STONEFLY —
Perla abdominalis

Stoneflies are to be found in a variety of aquatic habitats. Many of the smaller forms, which need well oxygenated water, occurring in fast-flowing streams, but the species shown here is found in slow rivers and lakes. It may be recognised as a nymph by its flattened shape with two long tail filaments and in the adult stage by the four wings, folded flat over the abdomen. Stoneflies are sluggish creatures spending much of their time under stones or in vegetation, and as adults flying rather laboriously, if at all. Their food varies with the species which in some cases are entirely vegetarian while in others, such as *Perla*, small aquatic animals are eaten also, during the nymphal life which may last for two years. At the end of this period, the insects crawl out of the water and on a stem or twig split their last nymphal skin and leave it, a grey ghost of their former selves. Sometimes these spent cases may be seen in very large numbers near to the water. After mating, the females lay clumps of eggs on the water surface. These sink to the bottom and separate before they hatch.

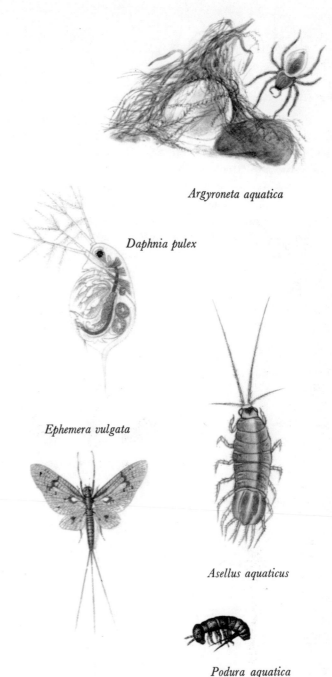

Argyroneta aquatica

Daphnia pulex

Ephemera vulgata

Asellus aquaticus

Podura aquatica

Perla abdominalis

BANDED AGRION —
Calopteryx splendens

(In America this insect is known as *Agrion splendens*). This damsel fly which is common throughout the summer months in southern Britain is known in mainland Europe from Scandinavia southwards and east into China as well as across America. It may be seen in open country where the pools and slow-flowing streams have a muddy bottom, for this is the environment in which the nymph passes its life. As in all damsel flies the nymph has three gill plates at the end of its body; these distinguish it from the immature forms of other insects. Like the adult it is a voracious carnivore, capturing many kinds of small creatures with its extendable jaws. The adults may live for several weeks, during which time they may often be seen resting on the floating leaves of water plants. The illustration shows a male; the females have metallic green bodies and unbanded wings.

BROWN AESHNA —
Aeschna grandis

One of the largest dragonflies found in Britain, where it is established in the south, is the Brown Aeshna, which measures about 75 mm. in length and 105 mm. in wingspan. It occurs also eastwards across Europe from France but not in the warmer parts of the south. The body is almost entirely tawny brown in colour apart from two yellow stripes on the side of the thorax and some blue spots on the abdomen, and in the male, blue eyes. The wings are yellowish in both sexes. The time of year at which the Brown Aeshna is to be seen is usually during late Summer and Autumn, but it is fairly long lived and is recorded among the insect species which may undertake long distance migration. It is often found well away from water, even in urban areas, although its more usual habitat is near to the banks of streams and pools with reedy borders. The female lays her eggs singly, in the stems or leaves of water plants, among which the nymph lives for two years before transforming into the adult form. The nymphs, which grow to a length of up to 45 mm., are, like the adults, entirely carnivorous.

CLUB-TAIL DRAGONFLY —
Gomphus vulgatissimus

The Club-tail Dragonfly, which is a common species in Central and Southern Europe, is rare in Britain, where it is known only from some areas in the south. It has a relatively short period of active life and is seen on the wing only in early Summer. The males fly rather slowly over the water, resting often on floating leaves, but until the time comes to mate, the females are often found among trees close to the water. The eggs are dropped into the shallow water and sink into the mud at the bottom where the nymph burrows for the three years of its existence. When the time comes for its emergence to the adult form it clambers out on to the bank, for its legs, although sturdy, are not suitable for gripping the stems of water plants.

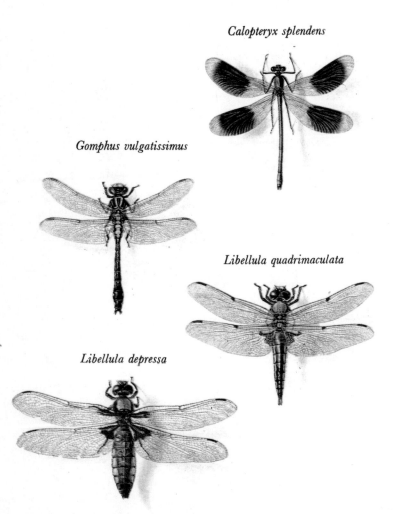

Calopteryx splendens

Gomphus vulgatissimus

Libellula quadrimaculata

Libellula depressa

FOUR-SPOTTED LIBELLULA —
Libellula quadrimaculata

This is one of the most widespread of dragonflies, occurring throughout Britain and Europe and much of Northern Asia and North America as well. Its size is rather variable but the body length is usually about 30 mm. It is on the wing throughout the Spring and early Summer and exceptionally may be seen as late as the autumn months. The eggs are laid in still water of all kinds including brackish pools near to the sea and peaty water on moors and heathlands. It may be seen darting from a perch near the water to snatch its prey which may include flies, butterflies or even wasps. The spread of the species is doubtless aided by its migratory habits, for it is one of the forms sometimes seen in huge swarms travelling long distances to suitable breeding areas.

BROAD BODIED LIBELLULA —
Libellula depressa

This dragonfly may easily be recognised by the size of its abdomen which in females may reach over 10 mm. in width. It is to be found throughout southern Britain and most of Europe, breeding in much the same sort of places as the Four-spotted Libellula. It is essentially a species of late Spring and early Summer, when it may be quite common locally, but in spite of this it is less frequently migratory than many other kinds of dragonfly. The nymph, which may be found in quite small pools, lives among the mud and plant stems where it stalks the small animals which are its prey.

COMMON CORIXA —
Corixa punctata

This bug is to be found throughout Britain and much of Europe in lowland ponds and slow rivers where the water is neutral or alkaline or only slightly saline. In some parts of Europe there are two generations in the year, but in Britain where there is usually only one, the adult insects first appear in midsummer. They swim strongly under water, but come to the surface to trap air bubbles which enables them to breathe when submerged. They fly readily and may be seen on the wing during the daytime. Their food may contain some small animals, but is mainly plant material, the digestion of which is aided by the secretion of special enzymes from glands in the head. The males make a creaking courtship song by rubbing part of their front legs, which carry patches of stiff hairs, against the side of the head. After pairing, the females, which live for some time longer than their mates, lay their eggs at night, attaching them to the stems of water plants. This may occur in the winter months but development of the nymphs does not begin until the water temperature has risen with the beginning of Spring.

BACKSWIMMER —
Notonecta glauca

This animal's odd name is descriptive for it does indeed swim on its back. Its size, about 20 mm. long, enables it to be seen easily, as it rows itself along with jerky strokes of its large hind legs in almost any body of standing water in the British Isles and much of Europe. Where it is absent it is replaced by closely related species. It carries a bubble of air which provides it with oxygen and which it must renew by periodic visits to the surface of the water. Here it often lies motionless but it is attentive to any disturbance in the water which may mean the presence of food. It will swim quickly towards any centre of movement but its eyes, although large, are not used until it has moved very close to its prey. This may be any of the larger invertebrates, small fishes, tadpoles or newts which are easily overcome by it. The powerful stabbing beak may be used by the Backswimmer to protect itself and care should be taken when handling these animals, for the bite can be quite painful. The male's courtship includes grasshopper-like

Nepa cinerea

Corixa punctata

Notonecta glauca

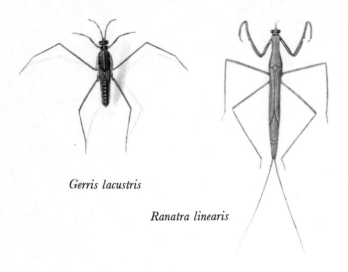

Gerris lacustris

Ranatra linearis

singing; the eggs are laid early in the year embedded in plant stems.

WATER SCORPION —
Nepa cinerea

This large bug, which measures up to 20 mm. long, is to be found in muddy, weedy ponds throughout Britain except for the extreme north, and also over much of Europe. Unlike the previous species, it is

Dytiscus marginalis

an inactive animal, living at the bottom of the water where it is camouflaged by its colour against the mud. From its hiding place it snatches its prey, which includes any passing small vertebrate and many kinds of invertebrate. This is held in its large, strong four legs in an embrace from which the smaller animal cannot escape. The long spike projecting from the rear end of the body is not associated with the Water Scorpion's predaceous habits and cannot harm anybody handling the creature, for it is a snorkel which when pushed to the surface enables the animal to breathe. Breeding occurs in the late Spring; the young develop through the Summer and overwinter in the adult stage. One of the curious habits of the Water Scorpion is its ability to feign death when handled and to escape when the attentions of an attacker are relaxed.

WATER STICK INSECT —
Ranatra linearis

The Water Stick Insect grows to a length of about 35 mm. It is closely related to the Water Scorpion which it resembles in its inactive way of life, for it normally sits among plants growing up from the bottom of rather deep pools, its long body and legs looking like scraps of dead vegetation. It is quick to seize any small animal which comes within reach and the adults are capable of taking prey up to the size of tadpoles. Under some circumstances the Water Stick Insect will swim, using its two hind pairs of legs and it may fly from one pool to another. In Britain it is to be found only in the southern counties. Like the Water Scorpion it may feign death if handled.

COMMON PONDSKATER —
Gerris lacustris

This is the most widespread water bug in Britain and Europe, occurring, often in large numbers, on stretches of still water of all kinds. As its name suggests, the Pondskater moves over the top of the water but does not sink through the surface film. This is possible because the feet are shod with a mat of dense hairs which trap an air cushion on which the animal keeps dry, although the water dimples under it. The middle pair of legs are used in unison, giving the animal a very jerky type of movement which is characteristic of it, while the hind legs are used as rudders. Pondskaters occur in a variety of forms, some having well-developed wings, others being short winged or wingless. No complete explanation is known for this, but it is thought that speed of development of the young may play a part as well as heredity. Pondskaters are carnivorous, feeding on any small animal which may be on the surface of the water. The struggles of a fly which has become trapped there will quickly alert any Pondskaters in the area which rush towards it and grasp it with the front legs and quickly suck it dry. The winged Pondskaters fly freely and usually overwinter some way from water, colonizing new pools early in the year. The first eggs are laid in May and are mature by late June when a second generation is produced. The behaviour of a whole group may be influenced by local conditions in the pool including the amount of plant growth and of light available at any time.

GREAT DIVING BEETLE —
Dytiscus marginalis

This fiercely predaceous beetle which grows to a length of 30 mm. is found throughout Britain, Northern and Central Europe and North America in pools of all sorts especially where there is plenty of weed cover. It swims well, using its broad hind legs. The illustration shows the male which is easily distinguished for he has smooth elytra while those of the female are furrowed. His front feet, unlike hers, carry circular discs, which are sucker-like and enable him to hold his mate during pairing. Although aquatic, it needs to breathe air, and does this by carrying a store under its wing cases. This has to be renewed occasionally and the beetle approaches the surface tail-first, opening breathing pores and raising its wing covers to take in more air which will be used when it is submerged. The eggs are laid individually in slits made in water plants in late Spring. The larvae are also carnivorous, attacking every small animal, vertebrate or invertebrate, which comes their way. The larva cannot chew its food, but it pumps a digestive

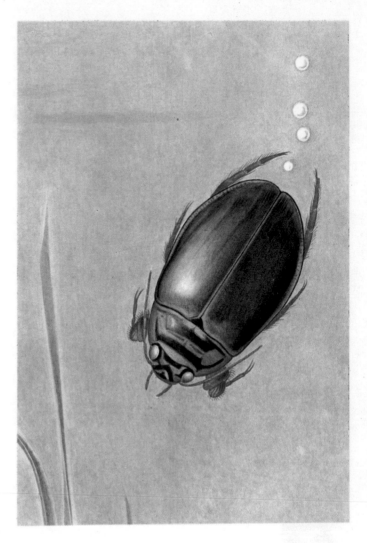

Acilius sulcatus

juice down narrow channels in its mandibles and when this has done its job of breaking down the tissues of the prey, the Great Diving Beetle larva then sucks up its predigested meal. Like the adult the larva has to breathe air, which it takes in through large breathing pores at its rear end. Its pupates in the bank of the pool and after emergence can fly strongly as well as swim.

LESSER DIVING BEETLE —
Acilius sulcatus

The Lesser Diving Beetle, which grows to a length

Hydrophilous piceus

Gyrinus natator

Donacia crasipes

Tabanus bovinus

of up to 18 mm. leads a carnivorous life like that of the previous species, but confines itself to smaller prey. None the less, if it invades a fish hatchery, it can do considerable damage to the young fry. It is widespread in Europe and Northern Asia, preferring weedy pools where it can easily stalk its prey and escape from larger enemies.

WHIRLIGIG BEETLE —
Gyrinus natator

In the summertime the surface of pools where there is not too much vegetation may be dimpled with the movement patterns of Whirligig Beetles. A number of species, all very similar to the one illustrated, occur through Britain and most of Europe, spreading into Turkey and Northern Asia.

The insect whizzes across the surface of the water, using its paddle-like middle and hind legs, but if it is alarmed can dive below the surface and remain submerged for some time. Like the Pondskaters groups of Whirligigs inhabit the same small area, feeding on small animals on or below the surface of the water. The upper facets of their eyes are adapted for seeing clearly in the air; the lower are for vision in water. Their success as predators and at escaping enemies is doubtless due to this, for they are both abundant and difficult to catch. The larvae, which live submerged are also predacious.

GREAT SILVER WATER BEETLE —
Hydrophilous piceus

This is the largest water beetle found in Britain, reaching a length of 45 mm., although it is becoming increasingly rare as many of the thickly vegetated ponds which are its favourite habitat are destroyed. It is to be found throughout the colder parts of the Old World, extending as far south as north India. It may be recognised, apart from its size, by the silver sheen of its wing cases and the fact that when rising to the surface for air, it does so head first rather than tail first in the way of the carnivorous water beetles, described above.

Although a skilful swimmer, using its second and third pairs of legs, which are flattened and fringed with hairs, it is not as fast moving as the carnivores. It feeds mainly on water plants, although it may sometimes turn to inanimate animal food such as frog spawn or to scavenging dead fish. The eggs are laid attached to floating leaves. The larvae are predacious, feeding on Water Snails, but they do not have the hollow mouth parts of the immature *Dytiscus* so they cannot pre-digest their food. Pupation is in the bank of the pond and the adults emerge and re-enter the water in the late Summer.

REED BEETLE —
Donacia crasipes

An insect often found in ponds and slow-flowing streams where water lilies abound is the Reed Beetle which may be seen resting on the leaves of water plants. Most of its relatives are live on land only but several species including this one have become adapted to aquatic life throughout Britain and Europe and much of northern Asia. The females lay their eggs on the underside of leaves and the fat white larvae crawl down into the water. They obtain the air which they need from plant stems, perforating the air channels and tapping them with a syringe-like process for the purpose. The pupae are similarly supplied with oxygen from the plants, which enables them to survive, but the adult beetles leave the water as soon as they emerge from the pupal case.

ALDER FLY —
Sialis lutaria

Alder Flies may be seen in early Summer in the vicinity of water in Britain and much of Europe. This species, which is the commonest British form, is cigar-brown in colour and about 25 mm. long and may be recognised by its heavily veined wings which are folded to make a ridged roof over the back while the insect is at rest. It spends much of its time resting and flies reluctantly.
The females lay large batches of eggs in clusters on stones or plants near to the water, and the larvae which hatch out in about fourteen days make their way to the water where they spend

about two years before they emerge as adults. Although they can swim by undulations of the body, their habitat during this time is the mud at the bottom of the pool, where they hunt any small creature they may encounter. They may be easily distinguished from other animals at the bottom of the pool by the single point at the end of the body and the tracheal gills. These are fan-like structures which at first might be mistaken for legs but which are in fact used for breathing.

CADDIS FLY —
Phrygaena grandis

Caddis Flies are common insects, in all non-polluted waters, but the many species show considerable specialisation to different ways of life. Some forms are associated only with fast moving mountain streams, others with quite different habitats. *Phrygaena grandis* is found in weedy ponds and ditches. It is the largest British species and has a larval length of about 50 mm. and an adult wingspan of 70 mm. The adults which are active mainly at night are at first sight like dull-coloured moths in appearance, but the wings are covered with fine brown hairs and the mouth parts are quite different from those of the moths. Although the adults are rarely observed, the larvae are among the best known of water animals, for the helpless creatures usually cover themselves with an armour of fragments of material made into protective cases. Each species has its own pattern; *Phrygaena grandis* uses pieces of leaves with which it forms a neat tube. Others make their protection of sand grains or plant fragments of various kinds.

HORSE FLY —
Tabanus bovinus

Horse Flies include some of the largest European flies. The species illustrated, which is known in Britain from damp areas in the New Forest and other ancient woodlands, may measure up to 25 mm. long and have a wingspan of 50 mm. It is generally the females which are seen for these have to have a blood meal to permit the ripening of their eggs. The males feed on nectar and so pass unnoticed as a rule. The Horse

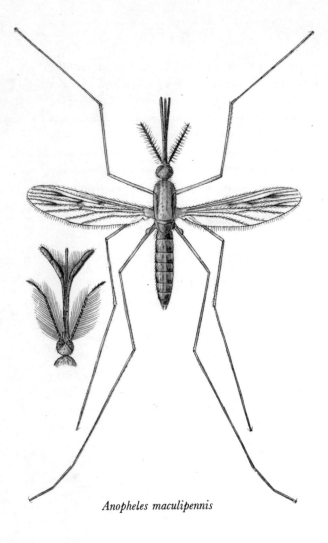

Anopheles maculipennis

Culex pipiens

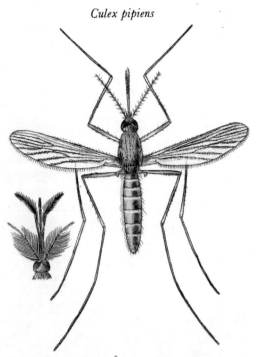

Fly usually preys on cattle, horses or deer, but human blood is equally suitable to her needs and will be taken quite readily. The fly approaches remarkably silently and the first indication of her presence is the painful stab of her proboscis as she starts her meal. The eggs are laid near to water for the larva needs to develop in damp places, under rotting vegetation or in the soil where it finds the small animals on which it preys. Worms and insect grubs are probably its chief food which it bites with hook-like mandibles before proceeding to suck the body contents dry. In the colder parts of Europe, including Britain, there is only one generation a year, active in late Summer; in some warmer areas there may be several. Although the bite of the Horse Fly is unpleasant, in Europe it is rarely dangerous to man or to farm stock, but in some tropical parts of the world Horse Flies transmit virus and bacterial diseases.

ANOPHELES MOSQUITO —
Anopheles maculipennis

One of the major groups of the small long-bodied flies commonly refered to as Mosquitos, contains insects related to the specimen illustrated above left. It may be recognised most easily by its resting position in which the abdomen is held at an angle to the place where the animal is sitting. In the Culex (see next entry) the body is held parallel to the substrate. Males may be distinguished by their feathery antennae, while those of females are thread-like. As with Horse Flies Mosquito males feed on plant juices, but the females must in almost all cases have blood before they can lay their eggs. The very long, fine mouth parts allow the sucking up of blood from the victims, but before the meal starts, the insect pumps into her host a tiny drop of an anticoagulant to prevent the fine column of blood in the proboscis from clotting. This is the cause of the itching commonly associated with insect bites, but in some cases more serious results may occur, for *Anopheline* Mosquitos may carry malaria parasites which they pass with their anti-coagulant to a vertebrate host. This has now been practically wiped out in Europe, but for many centuries the ague — a local name for malaria — plagued much of Northern Europe, especially near the coasts where a subspecies of *Anopheles maculipennis*, which is particularly liable to attack man occurs. The eggs of *Anopheline* Mosquito are laid in water, where the larvae develop. They are important food for fishes and many other aquatic creatures.

COMMON GNAT —
Culex pipiens

Common Gnats concentrate on birds for their blood meals, only rarely attacking human beings. They often hibernate in buildings, however, particularly in outhouses in the dark or rarely used rooms. There are many differences between this species and the previous one. These start with the eggs, which are laid in rafts on still water, unlike the singly eggs of the *Anopheline*. In both cases, however, the eggs produce aquatic larvae, easily told apart by their position in the water, for the *Anopheline* lies parallel to the water surface, while *Culex* hangs head downwards from it. They both feed on micro-organisms which they strain from the water, using a pair of moustache-like feeding brushes, which are vibrated rapidly to produce currents converging on the mouth. Anything trapped in the hairs of the feeding brushes is removed by bristles on the jaws. *Anopheles* can twist its head and feeds on objects just below the surface

film; *Culex* feeds at a deeper level. In spite of being legless the larvae can move quite fast through the water by lashing the body from side to side. The pupa, which is also aquatic, looks like a tadpole with the tail curled under the head. Unlike most pupae, it is highly active, and can break away from the surface where its breathing tube is in contact with the air, to swim to safety if need be. Depending on conditions, several generations may be produced in a single year.

FISHES

PIKE —
Esox lucius

The Pike is found in standing waters and slow-flowing rivers throughout temperate Europe, Asia and North America. It is well known for its predacious habits and although it is a valuable game and food fish is not welcome where other species are being preserved. The eggs, which measure about 2,5 to 3 mm. are laid in shallow water in Spring, a large female spawning about half a million over a period of several weeks. They are adhesive and stick to the water plants and sometimes are left stranded as shallow water recedes after spring floods. At hatching the little

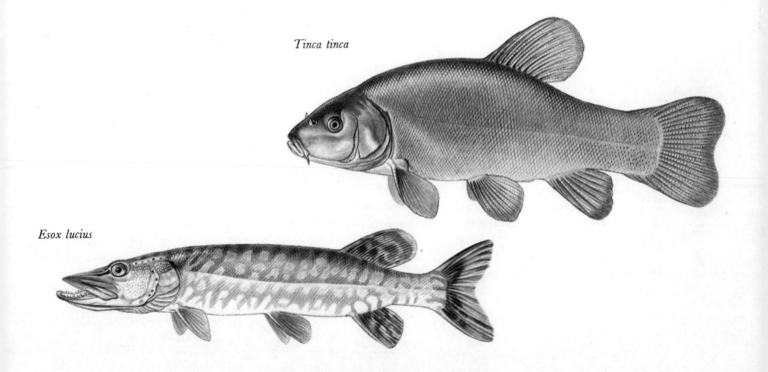

Tinca tinca

Esox lucius

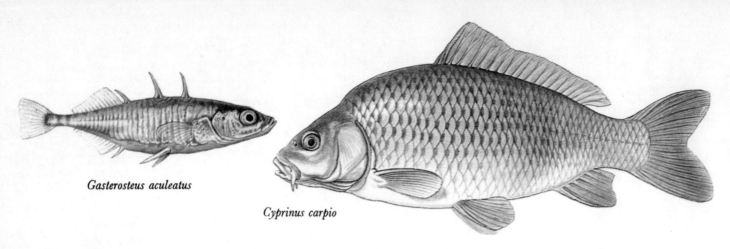

Gasterosteus aculeatus

Cyprinus carpio

fish remains attached to the water weed until the yolk sac is used up; it then becomes free swimming. At first it feeds on insects and other small water dwellers, but when about 40—50 mm. long graduates to feeding on other fishes. The rate of growth varies but is generally fast in the early years, and under favourable conditions the Pike may weigh 1 kilogramme in two years, though normally it would take twice this time. Males are mature at two years; females which grow larger, at four. Growth continues throughout life although it becomes slower as the fish ages. All very large Pike are females which may live to more than thirty years, at which time they may be over 1500 mm. long and weigh in excess of 35 kilogrammes.

CARP —
Cyprinus carpio

Carp were originally native to the slow streams of temperate Asia, but are reputed to have been introduced into European waters by the Romans. They now occur widely over much of Europe where the species is cultivated for food, and they have also been introduced into rivers in the United States. Many forms of Carp are known. Some are covered with normal scales, but the one illustrated is a Mirror Carp in which there are a few large scales only; Leather Carp have none at all. Carp thrive best in warm, weedy waters with a muddy bottom. They feed on a wide variety of foods, including insects and other small animals as well as plants of many kinds. Under ideal conditions growth can be very rapid and Carp may weigh 1 kilogramme in two to three years. They are long-lived fish and may survive for thirty years or more by which time they can weigh up to 30 kilogrammes and be over

1000 mm. in length. Large numbers of eggs are shed in shallow water in Spring. In warm areas the young develop rapidly and the males may be mature by the time they are three to four years old, but in the cooler parts of the Continent growth is slower and spawning may not be successful if there is not sufficient warmth.

TENCH —
Tinca tinca

The Tench, which is a shy creature living in weedy ponds and slow rivers, is to be found over most of Europe except for Greece and the far north. It can stand very low oxygen levels better than most other fish and can also survive in brackish water. During the summertime Tench feed on various molluscs and insects, but in the cold weather they become torpid, hibernating in the mud at the bottom of their pond or river. Spawning occurs in shallow water in early Summer, each female depositing clumps of eggs over a period of about two months, a large individual producing about 90,000 although few of these will survive. Growth is slow, and it rarely achieves a length of more than 500 mm. or weight of over 8 kilogrammes. Tench are valued as food in some areas and are cultivated in Tench ponds, particularly in the east of the Continent.

THREE-SPINED STICKLEBACK —
Gasterosteus aculeatus

Sticklebacks are to be found in ponds and ditches throughout Britain and in a broad band across Europe including most of France and southern Scandinavia, across to the Black Sea. Apart from this they occur as a species of coastal and esturine waters in all but the most southerly regions. In spite of their small size (for Sticklebacks rarely grow beyond 80 mm. long), they are aggressive and pugnacious, attacking all sorts of small animals for food. They are themselves eaten by many larger species of fish as well as birds such as herons and grebes and also by water mammals.

One of the Stickleback's main claims to fame is that it has been the subject of much intensive study into its breeding behaviour, which has given us an insight into a broad spectrum of behaviour beyond this species. The males, during the breeding season, develop bright pink colours; their undersides glowing water-melon pink and their eyes electric blue. They take up territories and drive away all rivals. Within its territory each male builds a nest of fragments of water weed and after one or more females has laid her eggs there the male fertilises and guards them and aerates them by spraying water over them. After they hatch, the young fry are cared for for some time by their father before becoming fully independent.

AMPHIBIANS

CRESTED NEWT —
Triturus cristatus

This animal is sometimes known as the Warty Newt, for the skin, although moist, is rough. Apart from Ireland, it is to be found throughout Britain, preferring ponds with rather deep water but plenty of weed

growth in which to breed. In mainland Europe, where it occurs from France to the Urals, several geographical races, differing slightly in size and coloration, are known. It is the largest newt species in Britain, with females occasionally growing up to 180 mm. long. The breeding season is in the Spring when the newts leave their hibernating places and return to the water. At this time, the male develops a high, jagged crest along the back and brighter colours than during the rest of the year. The courtship of the male newt involves constant movement of his tail and actions such as rubbing the flanks of the female with his snout. Finally he deposits a spermatophore, which is picked up by the female. The eggs, which are fertilised internally, are laid singly on the leaves or stems of water plants. About 200—300 are produced by a female in any one season. The tadpoles are full grown and matamorphose by the end of the Summer when they leave the water. They live in damp and mossy places, feeding on small invertebrates.

SMOOTH NEWT —
Triturus vulgaris

The Smooth Newt is the common newt species over Britain including Ireland and much of Europe, where it is to be found from Scandinavia and France across the landmass to Siberia. In the southern parts of the continent of Europe forms slightly different from the typical subspecies occur. It is smaller than the Crested Newt, reaching a maximum of about 110 mm. and in the breeding season the male can be distinguished by

Triturus cristatus

Triturus vulgaris (young)

his undulating crest, which is continuous from the neck to the tail-tip, although this is lost leaving only a slight ridge at other times of the year. The breeding places of the Smooth Newt includes almost any stretch of water in lowland country, where it will inhabit pools too small for the Crested Newt. During courtship the male faces the female, his tail, with which he is lashing the water, is bent back parallel with his body. A fully mature female lays up to 350 eggs some days after mating.

After the breeding season the newts leave the water and the males loose their crests and bright colours. The young are not mature until their third year and continue to grow beyond this time. They feed on many small invertebrates and are themselves the food of many predators, including carnivorous fishes such as pike, birds such as herons and a wide range of flesh eating mammals. During their tadpole stage they may be the prey of Water Beetles or their larvae, Dragonfly nymphs and a number of other predators both vertebrate and invertebrate.

FIRE BELLIED TOAD —
Bombina bombina
YELLOW BELLIED TOAD —
Bombina variegata

The Fire Bellied Toad and the Yellow Bellied Toad are closely related species which do not occur in Britain but are found widely on the continent of Europe. The Fire Bellied Toad is an animal of lowland regions and occurs from Denmark and south Sweden eastwards into Asia Minor and the Urals. The Yellow Bellied Toad is an animal of hilly and mountainous regions and is found from France through Germany south and eastwards into Greece. Both of these species are small, growing to not more than 50 mm. and both are essentially aquatic, spending most of their time in water, occupying tiny pools which are barely more than puddles if nothing bigger is available.

In all kinds of toads the skin contains poison glands which are the only defence that that these small weak animals have against strong predators. The poison produced has a strong effect on the eyes and soft parts of the mouth of any creature which attempts to seize them, although they are entirely harmless to human beings for the skin on our hands is too thick and heavy to be hurt in any such way. In these two species the glands are particularly well developed and their bright colours help predators to remember the toads' unpleasant characteristics, so the little amphibians are usually left severely alone. While they are swimming the bright colours can be seen fairly easily by other creatures in the water. When they are on land they are camouflaged by the dull brown of their backs, but if threatened they display their bright warning colours by arching their backs and holding out their legs.

During the breeding season both of these toads are vocal, the male Fire Bellied Toad having throat sacs which give its voice greater power. Males of both species develop horny pads on some of the fingers; these are more strongly developed in the Yellow Bellied Toad, which carries them on some of its toes as well. Spawning takes place from April onwards and in the course of the Summer a female may produce several batches of eggs.

GREEN TOAD —
Bufo viridis

The Green Toad is an Eastern European and Central Asian species found in very few places west of the Rhine, but it ranges from that river into Mongolia and from Scandinavia to North Africa. Apart from forest areas it is likely to be found in a wide range of habitats, even extending to brackish waters and can stand drier conditions than many other amphibians.

It is a fairly small species, rarely reaching as much as 90 mm. in length, characterised by the rather narrow head and the short web between the toes. The male has very large vocal sacs and in the breeding season which begins in April and may last well into the Summer, his trilling song may be heard. The female lays two long strings of spawn, which contain between 10,000 and 20,000 eggs. After breeding they leave the water, but are active mainly at night, sheltering during the daytime under stones or in holes in the ground.

NATTERJACK —
Bufo calamita

The Natterjack is closely related to the Green Toad and replaces it in western Europe, although the two species overlap in Denmark, western Germany and Poland. In Britain it once occurred in the London area but it is now many years since it has been seen there. It still survives in a number of localities, but these are widely scattered, mostly in sandy areas, including some by the coast. Here the Natterjack can dig itself into the ground for shelter during the daytime or for winter hibernation.

It is a rather small species, averaging about 60 mm. long, easily recognisable by the bright yellow stripe along the midline of the back and its short hind legs, which means that instead of hopping, it runs, mostly in a series of short spurts. The males have huge vocal sacs and their voices may be heard over long distances in the breeding season which starts in late March and may continue well into the Summer. Natterjacks do not appear to be as particular in the choice of a special pond as the Common Toad, and spawn may be laid in any shallow water within the area, including pools by the shore which are quite brackish. Development of the tadpole can be rapid and under ideal conditions may be completed within a month of hatching.

EUROPEAN TREE FROG —
Hyla arborea

This pretty little frog, which measures a mere 50 mm. in length, is to be found throughout Central and Southern Europe, but is quite absent from the northern areas. Attempts have been made from time to time to introduce it into Britain, but these have not been successful, although it is said to have survived for some years in the Isle of Wight, when a colony was founded there.

Adhesive discs on the fingers and toes enable the Tree Frog to climb well, and outside the breeding season it spends most of the Summer in bushes and trees, searching for the small invertebrates on which it feeds. It is camouflaged by its colour, which can change to match its surroundings from the bright

Bombina variegata

Bombina bombina

Bufo viridis

Bufo calamita

Hyla arborea

Rana esculenta

green shade illustrated to browns and greys. During the breeding season which lasts from Spring to early Summer, the males which have large vocal sacs are very noisy, and their croaking song may be heard especially on warm damp evenings or after rain. The spawn is laid in clusters about the size of walnuts in the edge of weedy ponds. The tadpoles may be recognised by their pale greenish-brown colour and their sharply pointed tails. A superstition that Tree Frogs can foretell the weather leads to their being kept in some places in glass jars. Although Tree Frogs can survive in captivity they need better conditions than are usually afforded them, and in any case as weather prophets they are extremely fallible.

EDIBLE FROG —
Rana esculenta

This frog, which occurs in Europe from southern Scandinavia and over much of Eastern Europe with the exception of the Balkans, has been introduced into Britain on a number of occasions and a few colonies still exist in south east England. Its close relative, the Marsh Frog, considered by some people to be a variety of the same species is now established in parts of southern Kent and Sussex. Both forms are strongly aquatic, but enjoy basking in hot sunshine although they are extremely wary and hide among weeds in the water if danger threatens. Both are voracious predators feeding on worms, snails and a wide range of other invertebrates. Both are noisy, especially during the breeding season, but the croaking song continues to some extent throughout the summer months. Spawning is usually in late May, and takes place in shallow water at the edge of the pool. In Britain, where the species is at its geographical limit

it is probable that only a very small proportion of the young survive.

REPTILES

GRASS SNAKE —
Natrix natrix

In many European countries the Grass Snake, which is widely distributed throughout the Continent, is called the Ringed Snake, in reference to the broken band of pale, usually yellowish, colour behind the head. In Britain it is widespread, apart from Scotland and Ireland, where it does not occur. It may be found in many habitats but prefers marshy places near to water. Grass Snakes swim well and frequently hunt their food under water, where they are able to stay submerged for some time without coming to the surface to breathe. Frogs, newts and fishes are the main prey of these voracious snakes, but other animals including mice, other reptiles, slugs and birds' eggs have been recorded among their food. When small they have many enemies and so are very wary; as they grow their size is to some extent a protection as is their ability to void a foul smelling fluid from the vent and this discourages aggressors very effectively. If badly

Natrix tessellata

Natrix natrix

frightened they may sham death and this apparently fools other predators into not attacking them. Mating usually takes place in the springtime and the females lay 30—40 eggs in midsummer. These soon hatch and by the Autumn the little snakes are about 160 mm. long. They are long-lived animals and an old Grass Snake may grow to over 1,000 mm., the biggest specimens coming from the warmest parts of their range.

DICE SNAKE —
Natrix tessellata

The Dice Snake is a large, slender species, reaching a length of 1500 mm. in Southern Europe. It is found from south west France across to Central Europe and Asia and is also known from North Africa but not from Britain. It lives by clear water, both flowing and still, and is to be found near the sea as well as by lakes in mountain regions. Its prey is chiefly fishes and amphibians which it catches in the water. In midsummer the females leave the water to find a safe place to lay their eggs — up to about twenty-five in number — usually deposited under leaves or in loose soil or rotting wood. Dice Snakes hibernate on land through the winter months, often using rodent burrows for the purpose. They are trapped in some areas for their skins which are used to make fancy leather goods.

BIRDS

REED BUNTING —
Emberiza schoeniclus

In the summer months Reed Buntings are to be found throughout almost the whole of Europe; in Winter the birds migrate southwards from the coldest parts of the Continent. The male, which is figured, is the more distinctive of the pair; his mate lacks his black head and bib but has instead dark brown cheeks surrounded by white and a buffish white bib, outlined in brown. Both birds measure about 150 mm. long. Their favourite habitat is a pond side, where there is plenty of thick cover, from which the male sings

Emberiza schoeniclus

Remiz pendulinus

his rather simple courtship song, and where, on or near the ground, the nest is made. Here the female lays four to six eggs which she incubates with help from her mate. The young, when they hatch after twelve to fourteen days' incubation, are fed by their parents on insects and other high protein food; adults feed chiefly on the seeds of various water plants. After the breeding season, during which two broods may be reared, Reed Buntings congregate in small flocks, searching open farmland for food.

PENDULINE TIT —
Remiz pendulinus

The Penduline Tit is a species of Southern and Eastern Europe, where it is found in thick vegetation beside ponds or in the overgrown banks of rivers or dykes. It measures only about 110 mm. long and is often difficult to see in the dense herbage where it spends most of its time, but it may be recognised by its sweet, drawn-out tsi-ee song. Its nest gives the bird its name, for it hangs like a flask with a side entrance tunnel from the outer branches of small trees such as sallows, or is sometimes constructed among reeds. It is warmly built largely of the cotton wool like parachutes of the seeds of poplars, aspens and willows. The walls may be more than 25 mm. thick, but this makes a snug nursery for the young. These are cared

97

for by the female alone, for although the male starts the building of the nest, he is aided in this by his mate, whom he subsequently deserts, building another nest and mating with another female. The eggs are white in colour and are incubated for fifteen days and the young are cared for for another fortnight by their mother who brings them a constant supply of small insects and spiders. Penduline Tits seem to be increasing their range towards the north and west but have not yet reached Britain.

GREAT REED WARBLER —
Acrocephalus arundinaceous

The Great Reed Warbler is a migrant bird which breeds in suitable waterside places throughout Europe, except for Britain and Scandinavia. It may be recognised by its large size — 190 mm. long — its long beak, clear eye stripe and its loud strident song which is often heard at night as well as during the day. This is sung from a perch in the reeds for it is less secretive than most of its relatives. It may nest colonially, building a deep cup of woven grass suspended in the reeds. Both parents incubate the four to six blotched eggs and feed the young for a further fourteen days on insects, spiders and snails. In some parts of Eastern Europe the Great Reed Warbler is heavily parastised by Cuckoos, and a high degree of resemblance between the eggs of the host and those of the parasite has evolved. Strangely, however, in other parts of its range it is not particularly attacked by Cuckoos.

REED WARBLER —
Acrocephalus scirpaceus

Reed Warblers are small birds, measuring about 125 mm. long, which are to be found in reed beds through most of Europe, being absent only from Scandinavia and the extreme north and east of the Continent. In Britain it is present only in the southern half of the country. In the field it is difficult to identify with certainty, unless it is singing, when its melodious song in which phrases are repeated several times, is distinctive. In flight the spread tail is characteristic. The cup-shaped nest is woven into the reed stems, several often grouped together. Incubation is shorter than that of the Great Reed Warbler, lasting for about twelve days and the young spend a further twelve days in the nesting stage. By October all the Reed Warblers have departed for Africa.

FIELDFARE —
Turdus pilaris

The Fieldfare, sometimes called the Blueback, is a member of the thrush family which nests, often in colonies, throughout the far northern parts of the Continent and in a broad area eastwards from central France. It is one of a group of species of birds which at the moment seem to be expanding their breeding range and in recent years this has extended to northern Britain, where small colonies now spend the summer months. The nest, which is lined with mud, like those of other thrushes, is usually built in small trees often close to the ground and frequently, although not always, near to water. The young are reared chiefly on insects and other invertebrates. During the Winter it migrates to warmer areas beyond its breeding range and at this time may be seen throughout Britain and in Southern Europe. It is a large bird measuring about 250 mm. long, strongly social and noisy when feeding on the berries of rowans, hawthorns etc. which are the favourite winter diet. It sometimes feeds in open country, often in company with the Redwing, another migratory thrush. Flocks of Fieldfares in flight are often strung out over long distances unlike the compact flight-grouping of most perching birds.

MARSH HARRIER
Circus aeruginosus

In the summer months Marsh Harriers may be seen over much of Europe except the extreme north, although they are also absent from all but the south east of Britain. They need large damp areas in which to breed, the large nest being made on the ground in extensive reed beds. Marsh Harriers may be recognised in flight by their long wings and tail and large size. The females are about 550 mm. long and have a wingspan of 1225 mm.; they are somewhat larger than their mates. In the early part of the breeding season the birds have acrobatic display behaviour which includes looping the loop, but in normal flight they flap the wings rather slowly for a few beats and then glide with wings held in a shallow 'V'. In this way Harriers quarter the ground low over the reeds, looking for the frogs, voles and young birds, such as gulls, which are their prey. The young are cared for by the female alone, but during the period of incubation and brooding the male provides all the food for his mate and family. The female leaves the nest only to take food from him, but apart from this remains strictly on guard. After fledging the young remain with their parents for some while. At the end of the Summer they migrate to their winter quarters in Southern Europe and Africa.

OSPREY —
Pandion haliaetus

Ospreys are large birds of prey with a winsgpan of up to about 1600 mm. They are found in isolated areas in Northern and Eastern Europe and North America throughout the summer months. In Winter they migrate south, the European birds mostly to Africa. They nest in tall trees, normally near to water, for the Osprey feeds on fish. It scans the water from a height of about 10 metres and plunges in feet first when it sees suitable prey. It emerges carrying the fish in its talons which are specially adapted to deal with the slippery, struggling victim. It may fly some

Acrocephalus arundinaceous

Acrocephalus scirpaceus

Turdus pilaris

Pandion haliaetus

Circus aeruginosus

Ardea purpurea

Nycticorax nycticorax

Ardea cinerea

distance to a suitable tree to eat the fish which it carries head first to reduce wind resistance. Ospreys are birds which have suffered from pesticide residues contaminating their food over much of their range. In some areas they have become much rarer in recent years. Before this, however, they were persecuted mainly in the name of fish preservation and in Britain they were exterminated as a breeding species before the end of the last century. A few Scandinavian birds had continued to use the migration route over Britain, and in the 1950s a pair bred successfully in Scotland. As a result of the most intensive protection the number of Ospreys breeding in Scotland has now increased although they are still not common birds. The publi-

city which the return of these birds attracted was turned to conservation use. The Ospreys have become a tourist attraction and road signs in the Scottish Highlands direct the travellers to the Ospreys. They are watched from a hide at a distance which will not disturb them by thousands of people each year.

GREY HERON —
Ardea cinerea

The Grey Heron is the most common member of the family in Europe, being present as a breeding bird in a broad area from Britain eastwards and migrating from the colder parts of this range to Southern Europe in the wintertime. It is a large bird, with a length of about 1000 mm. and a wingspan of about 1550 mm. It may be recognised in flight by the rounded wings, the slow heavy wingbeats, the long trailing legs and the position of the head which is tucked into the shoulders, not extended as with the cranes. Herons feed on a wide range of water animals but fish form the major part of their diet. When fishing they stand motionless in shallow water watching for unwary fish to approach. These are then snapped up by the long beak. Herons breed socially and the bulky nests made of sticks lined with fine vegetation are placed in tall trees. These long legged birds look incongruous in such a situation, but here the young are reared. Incubation, which is by both parents, starts with the laying of the first of the blue-green eggs so that one chick, as with the hawks and owls, is always more advanced than the rest. Later in the season groups of Herons may be seen fishing along the same stretch of river or lake-edge. In hard Winters Herons belonging to sedentary populations as in Britain often starve when rivers and lakes are frozen for long periods.

PURPLE HERON —
Ardea purpurea

Apart from a small area of Holland, the Purple Heron is found in Southern and Eastern Europe and occurs in Britain only as an occasional visitor. It is a smaller creature than the Grey Heron, measuring about 775 mm. in length and 1125 mm. in wingspan. The neck is relatively longer than that of the Grey

Heron and when at rest it is held in a strongly bent position while in flight it is curved to lie below the general line of the body. The trailing legs and feet are more obvious too. Purple Herons rarely perch in trees for they are mainly to be found in marshes and other wet areas where there is plenty of dense low growing vegetation. The nesting colonies are usually in reed beds, often in company with other species of ground nesting herons. At the end of the Summer they migrate to their winter quarters in Africa.

GREAT WHITE HERON —
Egretta alba

This beautiful member of the heron family is found almost throughout the world from South East Europe into Asia, across to North China and Japan. Closely related forms occur in tropical Africa, India, Australasia and New Zealand and it is widespread in temperate and tropical America. However, it does not live in North West Europe, and is recorded from Britain only as a rare vagrant. It is recognisable by its large size (890 mm. length and 1550 mm. wingspan) and pure white plumage — only the Little Egret which may occur in the same habitat could be confused with it and there are many differences between the species. The easiest to detect is probably the slender appearance of the Great White Heron compared with the more compact build of the Little Egret. Further differences between the two include the lack of a long crest, the black feet and the bill yellowish at the base and black at the tip. The nests are usually widely spaced in colonies not associated with other species. They are sometimes in small shrubs or trees but more usually in tall reeds, often in areas where there are several metres of water. There are three or sometimes four eggs which are pale blue in colour and are brooded by both parents for twenty-five to twenty-six days. Each pair rears only one family in the season but the breeding period is a long one since some birds start later than others. In spite of the reed nesting habit the Great White Heron is not specifically a reed bed bird and may be seen feeding at the edge of many kinds of open water, like the Grey Heron. Food varies with the season. Fish, aquatic insects and larvae, small mammals including mice and voles, reptiles such as lizards and probably young birds are taken as well as fresh water molluscs and worms.

BITTERN —
Botaurus stellaris

Another inhabitant of large areas of reed beds is the Bittern which is widespread as a breeding bird over most of Europe south of Scandinavia, but in Britain it occurs as a resident only in East Anglia. In spite of its large size, 760 mm. in length with a wingspan of 1550 mm., it is an extremely elusive bird and rarely seen, even in areas where it is comparitively common. When alarmed it assumes an elongated upright position, stretching its neck and pointing its bill skywards so that its striped colouring blends almost completely with the background of reeds. The presence of Bitterns, however, is indicated by the foghorn-like call which may be heard most frequently at dusk during the breeding season. It is said to be audible for a distance of several kilometres and in many

Botaurus stellaris

places has been the foundation of superstitious beliefs in ghosts and hobgoblins. Bitterns feed on a wide range of aquatic creatures but their main food is eels which they catch in great numbers.

NIGHT HERON —
Nycticorax nycticorax

The Night Heron is a summer migrant mainly to southern and eastern areas of Europe where it breeds colonially in bushes and trees in reed beds. The adult bird is easily identified by its rather thickset shape and black, grey and white colouring. Juveniles have a brown plumage but can be distinguished from Bitterns by their heavily spotted appearance. As their name suggests, Night Herons are largely nocturnal in activity but they may sometimes be seen at dusk or in the early morning, fishing for the small aquatic creatures which are their food. They often feed a considerable distance from their breeding place and may be seen commuting between the two areas flying in lines like flocks of sea birds. In flight the wingbeats are more rapid than those of other members of the family, a characteristic by which they may be identified when other features are indistinguishable because of lack of light. The young, which as in other herons hatch at different times because incubation starts with the laying of the first eggs, scramble out of the nest if disturbed and try to take shelter in the dense bushes.

LITTLE BITTERN —
Ixobrychus minutus

This bird, which is the smallest member of the heron family in Europe, occurs in Britain only as a passage migrant and does not breed here. It is, however, found in suitable locations throughout mainland Europe south of Scandinavia during the summer months. In some areas it may be common but it is shy and difficult to see in the dense reed beds in which it builds its solitary nest. It may give its presence away with its croaking call, which may be heard by day and night during the breeding season. If possible, it will avoid notice by freezing and will often run or climb to escape danger, but once flushed can be recognised by the dramatic contrast of buff and black on the wings of the male, and the characteristic flight with fast wing beats followed by a short glide low over the reeds. The Little Bittern feeds mainly on aquatic insects, but small amphibians and fishes are also taken. The nest is made on the ground and the young are cared for by both parents.

MUTE SWAN —
Cygnus olor

The Mute Swan is the largest European bird with a length of 1,520 mm. and a wingspan of 2,400 mm.

Anas platyrhynchos

Ixobrychus minutus

Anser anser

Anser fabalis

It is a resident species throughout Britain and part of North Central Europe and in many other areas occurs under conditions of semi-domestication. It is easily recognised by its large size and the graceful curve of its neck, by its totally white adult plumage and the orange bill with, in the male, a conspicuous black knob at the base.

In flight the powerful wingbeats produce a loud whistling noise, which can be heard for some distance. Swans' diet is almost entirely vegetarian, and the birds often upend to reach succulent leaves under water.

The nest is made of a pile of vegetation at the water's edge, on the shores of ponds, lakes or slow-flowing rivers and canals. The clutch of up to nine white eggs is incubated almost entirely by the female and although the male relieves her for short periods his main occupation is to guard the general area of the nest. At this time he is extremely aggressive to any intruders including human beings. The cygnets, which are covered with grey down on hatching, take to the water immediately, attended by both parents. They retain a dingy plumage until their second year when they grow their white adult feathers. In Winter, herds of swans may be seen in open water including the edge of the sea.

BEAN GOOSE —
Anser fabalis

The Bean Goose which may sometimes be seen on inland lochs in Scotland is a bird found throughout the high tundra of the Arctic areas of the Old World. It nests here although it migrates as far south as the Mediterranean during the winter months. It is a cautious and wary bird feeding mainly on water plants although in mainland Europe flocks are said to feed on stubble fields, a habit which is rarely observed in this species in Britain.

GREY LAG GOOSE —
Anser anser

This large bird with a wingspan of about 1,630 mm., is the most widespread and common of European wild geese. It is present as a breeding species through much of Northern and Eastern Europe and in much of Northern Asia also. Until the early part of the last century, it bred in East Anglia, but in Britain is now resident only in a few areas in Scotland although large numbers migrate here during the wintertime. As with all geese they are strongly social, breeding colonially and remaining in family parties through the year.

The nest is made in damp meadows and the four to seven creamy white eggs are incubated by the female while her mate stands guard a short distance away. They feed on many sorts of vegetation, including grasses and the remnants of field crops such as wheat or potatoes. In some areas they are valued because they remove any waste which has not been harvested and at the same time fertilise the field with their droppings.

MALLARD —
Anas platyrhynchos

Unlike the geese, in which males and females are very similar to each other, among the ducks there may be great differences in appearance between the sexes. The one illustrated is a male or drake in his spring plumage. At other times of the year he ressembles his mate who is dressed less gaudily in subfusc browns and greys. Mallards are to be found throughout Europe and much of Northern Asia and North America. They are the least fussy of ducks as to their breeding place and have been recorded from, and as nesting by, swampy places, pools and slow-flowing streams of all sorts. They adapt well to human company and are often found in small and polluted ponds in the middle of towns.

The nest is made by the female, who incubates the large clutch of greenish-grey eggs for about twenty-six days while the male stands guard a short distance away. As soon as they have hatched the ducklings are led to the water by their mother. They are able to swim well and will dive to escape danger, although this is a habit which they very rarely use later in life, for Mallard are surface or dabbling ducks and get their food, which is mainly vegetation of various kinds, by upending in shallow

Anas crecca

Anas querquedula

Spatula clypeata

Aythya ferina

Aythya fuligula

water. Outside the breeding season Mallard are social, living in flocks which may be very large. From early Winter onwards, however, the birds pair up, and may be seen as close sub-units within the flock, which may be found on any sort of open water including the sea.

GADWALL —
Anas strepera

Gadwalls, are to be found across Europe eastwards from Britain and occur also in Asia and North America, but are not common. The duck is similar to a Mallard, but slightly smaller with a length of about 510 mm. and a wingspan of about 890 mm. The drake is not so colourful as a male Mallard, his only bright colour being the fox-red speculum in the wing. They sometimes nest colonially in isolated marshy or lakeside areas, preferring places where the vegetation is dense. As with most ducks, the female incubates her clutch alone, which takes about twenty-seven days. At the end of the breeding season, Gadwall gather in small flocks, occasionally associating with other surface feeding ducks. They migrate southwards and westwards so that they may be seen throughout Britain in winter, although they are not common nesting birds. Their southerly movements take them as far as North Africa and the Caribbean.

TEAL —
Anas crecca

The Teal is the smallest European duck, measuring only about 350 mm. in length and 610 mm. in wingspan and is very common. Its breeding range includes all of Europe except for Spain, Italy and Greece, although its winter migrations take it into these areas. It is also found across Asia and into North America and the American subspecies, the Green Winged Teal, is sometimes seen on the eastern side of the Atlantic. Teal are often to be found on quiet small ponds. Unless they are in an area where disturbance from human activity is unlikely, Teal are active mainly after dusk, feeding on a variety of plants along with some molluscs and other invertebrates. If alarmed a flock of Teal may spring from the water and take off at high speed. They normally travel in a tight bunch rather than the chevrons or lines characteristic of most other ducks.

GARGANEY —
Anas querquedula

The Garganey, which is slightly larger than, and not as common as the Teal, is a widespread summer resident over much of Europe. Some occur in Britain, where they may breed, particularly in the south east. The kind of habitat favoured is a marshy area, with rank grass, where there will be plenty of food, especially vegetation of various kinds. Here the female makes her nest and incubates eight to twelve eggs for about twenty-three days. Like Teal, Garganey fly fast, but unlike them are usually found in pairs or small flocks only.

SHOVELLER —
Spatula clypeata

This bird gets its name from its broad beak, which is quite unlike that of any other species. It is to be found throughout the year in the British Isles and is a summer resident through much of Europe except for Scandinavia and the extreme south. From these areas it winters in Southern Europe and Africa. It also occurs in Asia and North America. Meadows near to water are favoured as nesting areas and here the dull coloured female incubates nine to twelve eggs for about twenty-four days. The food of Shovellers is almost entirely vegetation and tiny animals, which it sifts from the water by paddling rapidly through shallow water with its beak. At is does this the edible particles are trapped in the lamellae or comb-like structures which hang down from the upper bill.

PINTAIL —
Anas acuta

The handsome drake gives this species its name, for his tail feathers form a sharp, elongated point. This,

coupled with his dark brown head and rather long neck which is white at the front and sides, make him unmistakeable. The female, which is mottled brown like many ducks, should be recognised by her relatively long tail and neck. Pintails are found throughout most of Northern Europe including parts of Britain as summer residents, migrating south and west in the wintertime. Like the previous species it occurs in Asia and North America. It often associates with other species of duck, and is largely nocturnal in its activity.

POCHARD —
Aythya ferina

The Pochard is a diving duck and spends much of its time on the water, resting there by day and feeding at night. Unlike the species so far considered it can dive and swim and obtains a good part of its food under water, remaining submerged for up to half a minute. It is a large bird with a wingspan of about 785 mm. When it takes flight it cannot spring from the water like the dabbling ducks, but patters across the surface, beating its wings laboriously until it is airborne. It breeds in Britain and in a broad band across Europe south of the Gulf of Finland reaching Siberia. The nest is usually in reeds, but near to open water, and the large clutch of greenish eggs are incubated by the female alone for twenty-four days, during which time she never leaves them for more than a few minutes. The food, as with most ducks, is mainly vegetable but includes some small animals as well. There is a general southerly migratory movement in Winter although some slight movement may occur throughout the year.

FERRUGINOUS DUCK —
Aythya nytoca

This species, which is the smallest of the diving ducks, with a wingspan of only 661 mm., is found only in Central, Southern and Eastern Europe in lakes surrounded by dense vegetation. Both sexes have a rich brown coloured head, neck and breast, although it is a more intense and beautiful shade in the male. The underparts are white although this can normally be seen only in flight. Their behaviour and way of life is much like that of the Pochard.

TUFTED DUCK —
Aythya fuligula

This is a widespread species of the northern old world. In Europe it breeds from Britain almost to the most northerly limits of the continent and is found east into Siberia. In Winter it migrates south to the Mediterranean and beyond. It has recently been extending its range southwards and is now, after the Mallard, the commonest of ducks in Britain and Central Europe. They are to be seen on rivers, lakes and even the ponds in public parks in cities, where their jaunty appearance and their habit of diving and bobbing up suddenly make them amusing favourites with visitors. Their breeding places are secluded and the nest is usually in good cover near to water. The female, which is dark brown in colour and lacks her mate's long crest and obvious white side patches, incubates the eggs for about twenty-

Podiceps cristatus

four days, when the jet black fluffy ducklings hatch. Their food includes plant and animal material and is gathered chiefly under water.

GOLDENEYE —
Bucephala clangula

The Goldeneye is a duck of the far north of Europe, Asia and America, although it is seen on its winter migrations as far south as Italy. Breeding starts early and by March the Goldeneyes have reached the forested regions in which they nest. However the nuptual display of the males, which throw their heads back and swim towards their mates with their bills pointing skywards, may often be seen before this on the wintering grounds. In spite of the fact that they are extremely clumsy on land, the Golden-eyes nest in hollow trees, sometimes using holes well above the ground or even nest boxes, where these are provided. The female incubates the eggs for about thirty days, and then leads her ducklings, which are black with white cheek patches, to the water. If their nest hole is a lofty one they scramble out and jump confidently to the ground. Outside the breeding season Goldeneye are often seen at sea. They are excellent divers and can stay submerged for over half a minute, as they search for their food which includes plant and animal material.

RED CRESTED POCHARD —
Netta rufina

This species is found mainly in extreme South East Europe but small colonies occur across the continent and it may occasionally be seen as a winter migrant to south east England. It is possible, however, that a bird in this area would be one which has escaped from captivity for this handsome duck is often kept in waterfowl collections. Although technically a diving duck, it obtains much of its food, which is mainly vegetation, from the surface in the manner of the dabbling species. Its favourite nesting places are lagoons where dense growths of willows or sallow bushes give shelter. Here the female incubates her eggs for about twenty-seven days. Outside the breeding season they are usually seen in pairs or small groups.

GREAT CRESTED GREBE —
Podiceps cristatus

The Great Crested Grebe may be seen in the breeding season on lakes and ponds with dense fringing vegetation. In Britain some birds are entirely resident; others move to larger bodies of water, or the edge of the sea during the coldest weather. They are exclusively aquatic birds, rarely coming voluntarily to dry land and taking to the wing reluctantly, although they fly strongly once they are airborne. If alarmed the first instinct is to dive and swim some way under water where it has been recorded as remaining submerged for fifty seconds. In some areas they nest socially but in England a solitary pair will exclude others from their territory. Nesting is preceded by an elaborate mutual display. Both birds swim towards each other, heads low and ruffs extended and the ceremony includes head shaking and diving for water weed which is presented to the partner. There is also the so-called 'penguin display' in which the birds maintain an upright position breast to breast almost entirely out of the water. The nest is a floating platform of reeds and chalky white eggs are quickly stained dark brown on it.

BLACK-NECKED GREBE —
Podiceps caspicus

The Black-necked Grebe, which is only about 280 mm. long, is found through much of Britain and Europe south of Scandinavia and lives also in suitable areas of North America and Asia. The habitat is large ponds and lakes with surrounding reed beds, in which they nest usually in colonies, although outside the breeding season they are generally seen only in pairs or in small groups. Aquatic insects form the main food although little fishes may also be taken. The nest, as in all grebes, is a floating platform of reeds and the incubation is shared equally by both parents. Unlike the majority of birds in which incubation is delayed until the clutch is complete, in the Black-necked Grebe it starts with the first egg and a second brood may follow the successful rearing of the first.

Bucephala clangula

Netta rufina

Podiceps caspicus

Podiceps ruficollis

Charadrius dubius

LITTLE GREBE or DABCHICK —
Podiceps ruficollis

Measuring only 220 mm. in length, the Little Grebe is the smallest member of its family, but is found through all of Britain and much of Europe south of Scandinavia. It is also known from much of Asia and Australasia. It is a most elusive little bird, diving and bobbing up in the water many metres from the spot where it was last seen, but it is more catholic in its choice of living place than most Grebes and

107

Vannellus vannellus

in Britain it often colonises stretches of water formed in old quarries or gravel workings. The nest is the usual floating platform and the birds have the habit, common to all Grebes, of pulling a few strands of water weed over the eggs if they have to be left. In Britain it is a largely resident species; in Europe it migrates in a southwesterly direction during the wintertime.

LITTLE RINGED PLOVER —
Charadrius dubius

This small wader, which measures only about 145 mm. is to be found throughout Europe except for the far north and the greater part of Scandinavia but has bred only in the south eastern counties of Great Britain. In the late Autumn Little Ringed Plovers migrate southwards to Africa. It may be compared with its slightly larger relative, the Ringed Plover, but may be distinguished by the double white stripe across the face and the lack of white in the wings when the bird is in flight. Characteristically it is a bird of inland areas found by fresh water and often nesting on shingle banks in rivers or by the side of lakes. The four, strongly camouflaged eggs are laid in a shallow scrape in the ground, scantily lined with bits of grass. Incubation is by both parents, and the chicks are precocious and active soon after hatching.

LAPWING or PEEWIT —
Vanellus vanellus

Breeding throughout Europe except for the Mediterranean areas and the far north, the Lapwing is essentially a bird of farmland and the margins of fresh water, finding most of its food in areas of disturbed soil and nesting in meadows, moors and marshes. In Britain it may be seen as a resident and migrant, for birds in the northern part of the range move to the south in the wintertime. In flight they are easily recognised by their broad, boldly patterned wings and their habit of jinxing from side to side. In the springtime acrobatic displays precede mating and should there be any alarm the birds' rusty gate call speaks their alternative name. The rudimentary nest is on the ground and here the sitting bird is well camouflaged by the dull colours of its back. It is careful when approaching or leaving the nest not to do so directly, but slips away through the grass so that it will not give its position away. In spite of this Lapwings are easily disturbed, but will defend the nest area courageously, although crows and other marauding birds often succeed in stealing the eggs once the birds have been driven off by another creature — perhaps a wandering dog for example. Several pairs often nest in the same field but they are not truly social. The four eggs are incubated by both parents and the young are active shortly after hatching. After the breeding season large flocks may be seen feeding on farmland.In Britain Lapwings and their eggs are totally protected; in some parts of Europe they are highly regarded as gastronomic delicacies. The nasal call 'vee-veet' gives rise to the alternative name Peewit.

REDSHANK —
Tringa totanus

Redshanks are very commonly found in marshes, lowland moors and by the sea shore of much of Europe except the forested north and the species is also present in the cooler parts of Asia. Some birds in Britain are residents; others from colder areas migrate in Winter as far south as Africa. The Redshank is a restless, noisy bird and its musical trisyllabic whistle is often the first warning to other birds of the presence of human or other undesirable company. The males return to the breeding areas before the females, and here the fluttering display-chases may be seen. The nest, as with all waders is a shallow scrape in the ground, lightly lined with grasses. Four eggs are laid and incubated by both parents. Although the chicks are active as soon as they are hatched, they are cared for by the parents for some time. It is a month before they can fly and forty days before they are self sufficient, and can find the insects, worms, snails and other small invertebrates which are their food.

BLACK-TAILED GODWIT —
Limosa limosa

Well-watered areas in river estuaries, reclaimed grassland and marshes and mud flats are the home of the Black-tailed Godwit, a large wader which breeds from Holland and Denmark across Central Europe into Asia with outlying populations in Iceland and eastern England. The Icelandic birds migrate to southern Britain during the winter months; the European populations move to the Mediterranean borders and to Africa. They may be recognised by their long straight beaks, and in flight the large amount of white on their wings as well as the distinctive wide black bar on the tail. The four eggs are laid in a shallow scrape, thinly lined with grass and the male plays a major part in the incubation which lasts for about twenty-four days. The young are well camouflaged by their colouring and remain hidden during their early days, in tall vegetation near to the nest. Black-tailed Godwits feed on a wide variety of invertebrates, which they obtain by wading and probing mud which is under several centimetres of water. They often submerge their heads completely as they do so.

SNIPE —
Gallinago gallinago

The Snipe is a small wader, measuring only 250 mm. long, which is common in wet meadows and marshes. It occurs throughout Northern and Central Europe,

much of temperate Asia and North America, migrating to warmer areas during the wintertime. Britain and Western Europe have a resident population as well as receiving some overwintering birds from the far north. It is a wary bird but often relys on its camouflaged colouring for safety. If predators approach too closely Snipe rise rapidly from the ground and dart off with a zig-zag flight which is doubtless the origin for the collective noun for the species — "a wisp of snipe." Snipe in springtime have a display

Tringa totanus

Limosa limosa

Gallinago gallinago

Sterna hirundo

flight in which the male climbs to a height which may be over 100 metres and dives down vertically with tail feathers spread. The air thrums through them to produce a characteristic drumming sound. Four young are reared in a well-hidden nest, near to the wet, oozy places where the Snipe feed on many kinds of small invertebrate.

COMMON TERN —
Sterna hirundo

In England the Common Tern is a bird of the sea shore, but in Scotland and mainland Europe it is equally found by inland waters, where it nests in dense colonies. Terns in general are distinguished from other water birds by their long wings, forked tails and buoyant flight and their habit of hovering and diving for fish. The Common Tern, which frequently flocks with other kinds of tern, has more black on the underside of the wings and a brighter red bill with a black tip than most similar species. Common Terns nest on the ground with little protection for the eggs but these are blotched and so pebble-like in appearance that they are extremely difficult to see. The female does the major share of incubation but is fed by her mate during this time. The chicks remain in the nest for several days after hatching, brooded by their mother, but subsequently are fed by both parents on small fish, crustaceans, molluscs and insects.

BLACK TERN —
Chlidonias niger

Black Terns are small members of the family, measuring only 235 mm. long. They are known only as passage migrants in Britain, but they are widespread through much of Southern and Central Europe, temperate Asia and North America. Outside the breeding season they may often be seen along sea coasts. They breed, however, in wet marshy areas or in reedy meres making a floating heap of water weeds for a nest where incubation is carried out mainly by the female. The colonies are usually not large and the birds may change their location from one place to another with no apparent reason after long use of a particular site, but this is a habit which is found to some extent in all terns. Black Terns rarely dive for their food in the manner of the Sea Terns, but may snatch fish from the surface of the water. They feed also to a large extent on aquatic invertebrates and often hawk after flying insects.

BLACK-HEADED GULL —
Larus ridibundus

The Black-headed Gull occurs as a breeding species throughout Britain, Western, Central and Eastern Europe and is extending its range around the coast of Scandinavia towards the north. It is known in the southern areas of the Continent as a winter migrant. The name Black-headed is misleading for the head colour is chocolate brown and then only in the spring and summertime; outside the breeding season it is white with a dark brown spot remaining behind the eye. It is the most adaptable of the gulls, breeding colonially in many sorts of inland sites mainly in marshy areas, where the nest is protected against rats and other predators by water. Three eggs are normally laid and incubation is by both parent birds. The young are heavily camouflaged and difficult to see, lying still in the nest if danger threatens. Black-headed Gulls feed on a wide variety of foods which are mainly animal in origin and include worms, grubs and molluscs obtained from the country side often by following the plough, while in urban areas they scavenge from rubbish dumps. They have discovered that an easy living may be made in towns and in wintertime are to be seen mingling with the feral pigeons in London and other cities in Britain and in some large continental towns. They never become as tame as the pigeons and commute to out of town roosts at night, but the degree of their adaptation to land life may be seen as they perch, despite their webbed feet, on fences or flagpoles.

LITTLE GULL —
Larus minutus

Tussocks of grass in marshes and swamps in North Central Europe are the breeding places of the Little Gulls, which may be seen in Britain and round the

110

coasts of Europe as winter visitors from November to April. It may be recognised by its small size (only 250 mm. long) and its rounded wingtips, dark colours below the wings and grey above. In the breeding season the head and upper part of the neck are black, covering a much greater area of the bird than does the dark hood of the Black-headed Gull with which it is often seen. Food includes small fishes, insects, seaweed fragments and grain.

COOT —
Fulica atra

The Coot is absent from most of Scandinavia and the far north of Europe but is otherwise to be found as a breeding bird over almost the whole of the Continent, although it is present only in the summer months in the colder part of its range. It is distinguished from its near relative, the Moorhen, by its slightly larger size (about 360 mm. long), its white beak and frontal shield which has given rise to the phrase 'as bald as a Coot' and its rounded silouhette when swimming. Although they spend most of their lives on or near water Coots do not have webbed feet but their long toes carry broad lobes; and may dive after an upward jump. It normally requires a fairly large area of water for breeding and does not usually nest on a pond less than about half a hectare in extent. The nest is placed at the edge of the reed bed and made of a pile of stems and leaves of reeds and sedges. The brood may be a large one and as many as fifteen eggs have been recorded from a Coot's nest. Although they are quarrelsome and often aggressive to other water birds Coots may nest colonially and after the end of the breeding season form flocks on large stretches of open water.

MOORHEN —
Gallinula chloropus

Moorhens are to be found almost throughout the temperate world, occupying lakes, ponds and swampy areas. They are easily distinguished from Coots by the upturned tail which gives them a characteristic silhouette when swimming. Their toes are quite

Larus ridibundus

Chlidonias niger

Fulica atra

Galliula chloropus

Rallus aquatitus

Grus grus

unwebbed and as they swim with each stroke of their legs the head jerks forward as if the whole process were a considerable effort. Moorhens are often seen out of water, foraging in damp fields for the small animals, shoots and seeds on which they feed and although they are clearly more at home on the water, they sometimes perch in low trees. The Moorhen's nest is a deep bowl made of water plants, situated at the edge of the reed bed and is often decorated with leaves or other material, even litter such as paper bags round the outside. Incubation is for about twenty days after which a second brood is reared and this is sometimes followed by a third. The fluffy chicks can feed themselves after three weeks but are dependent for about five. When the parents are busy with a later brood it is quite usual to see chicks of the second brood being cared for by those of the first hatching.

112

LITTLE CRAKE —
Porzana parva

Coots and Moorhens are among the most easily observed of water birds, but their relatives the Rails and Crakes are extremely secretive and rarely seen. The Little Crake, which is a bird of Central and Eastern Europe is fairly common in some areas, but with its camouflaged colouring and small size (about 170 mm. long) it usually escapes notice. It migrates in April from Southern Europe to its breeding grounds in waterlogged, swampy areas with dense vegetation in which the nest is made, and up to eight young are reared. Its food is varied and includes grubs, worms, molluscs and seeds of water plants. The Little Crake's call is usually a series of barking notes followed by a trill.

WATER RAIL —
Rallus aquaticus

Water Rails, which measure about 265 mm. in length, are the largest of the European Rails, but they are none the less very rarely seen, for they are secretive birds living in swampy areas and pools with dense vegetation where they are most active at dawn and at twilight.

They are more frequently heard than seen, for they have a wide vocabulary of calls, many of them unmusical grunts and whistles which sound quite unbird-like. These are most likely to be given at night, if the birds are disturbed and may well have given rise to stories of water spirits.

Although they occur in Iceland, Water Rails are mainly found in the warmer parts of Europe, reaching only as far north as the most southerly areas of Scandinavia and migrating southwards after breeding in the colder parts of their range. The nest is usually hidden in a clump of sedges or reeds. The young leave it soon after hatching, although they are still in need of parental care. Water Rails are double-brooded and the older chicks, like Moorhens, often help care for the younger brood.

CRANE —
Grus grus

Cranes, which once bred in Britain, survived in East Anglia until the early 17th century but returned as winter visitors until the 1800s. It is now extremely rare for a wild Crane to be seen anywhere in this country. They would be difficult to mistake for any other bird for their large size (length about 1050 mm. with a wingspan nearly double this), long necks and legs which are carried outstretched in flight, and their bushy tailed appearance on the ground distinguish them from Herons and other wading birds. Their breeding grounds are waterlogged areas, with or without trees, in Northern and Eastern Europe and Asia. During the winter months they migrate to the southern parts of Spain and Italy and to North Africa. As a prelude to mating Cranes perform display dances during which they bow deeply with drooping wings and leap high into the air. The nest is usually near to water, built up above the level of possible flood and here both parents incubate the clutch, generally of two eggs, for twenty-nine to thirty days. Crane chicks have shorter beaks than their parents but as they become independent they develop the adult form. They feed on a wide variety of plants and invertebrates occasionally taking frogs and small reptiles or mammals.

MAMMALS

MUSK RAT —
Ondatra zibethicus

Musk Rats are natives of North America inhabiting streams and ponds from Alaska to Louisiana. They were introduced into Europe as captive animals farmed for their fur in the early part of this century. In almost all areas where they were kept, some escaped and the descendants of these animals have now established themselves widely throughout Europe, aided in some areas by deliberate releases. In Britain, accidental escapes found a haven in the ponds and slow-flowing rivers where they had no natural enemies. Like many rodents they breed rapidly and as

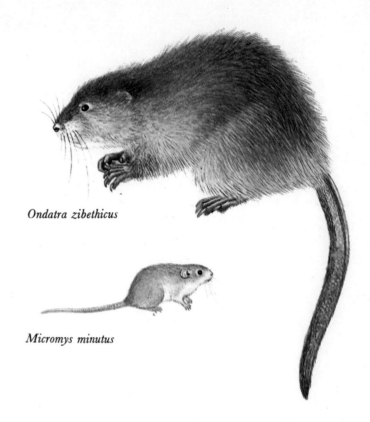

Ondatra zibethicus

Micromys minutus

a result the population increased and the Musk Rat, which has a body length of up to 400 mm. plus a tail length of up to 275 mm. began to do substantial damage to the environment. This was because of the amount of wild and cultivated plants which they fed on and the damage which their burrows did to the water systems of the country. Where the populations were really dense the deep tunnelling into the retaining banks of ponds and the sides of rivers led to their general collapse and the reduction of a well drained area to one of marshland. Escapes from Musk Rat farms in Britain began in the late 1920s. In the early 1930s the Ministry of Agriculture and Fisheries decided that even the addition of a potentially valuable fur-bearing animal to the fauna could not be tolerated at the expense of the damage done and so an all-out campaign was mounted to eradicate the Musk Rat from Britain. This was achieved after several years and the import of Musk Rats into the United Kingdom is now prohibited. In spite of this, in some European countries less heavily populated and more intensively farmed than Britain, the Musk Rats are valued for their fur.

113

HARVEST MOUSE —
Micromys minutus

The Harvest Mouse, which is Europe's smallest rodent measuring at most only about 148 mm. long, is to be found through most of Britain and Europe apart from Spain and the extreme south and far north. Its favourite summer habitat is in tall, rank vegetation, where it can get shelter and food and where its ball-shaped nest, woven among the plant stems, makes a safe undisturbed nursery for its young. In the old days, grain fields made an ideal home for the Harvest Mice. Now, with the increasing use of combine harvesters, the mice are destroyed as the crop is cut, and in Britain at least they have virtually disappeared from arable farmland. They are still to be found, however, in areas of undisturbed tall grass such as often grows in damp pondside areas, where they climb, using their tails in a prehensile fashion.

STREAMS
AND
RIVERS

INTRODUCTION

Most of Europe's great rivers rise in mountain regions and flow, gathering tributory streams, for a course often of thousands of kilometres to the sea, each major sea area receiving the water of one or more large drainage basins. Thus the Tagus, Garonne and Loire flow into the Atlantic, the Rhone to the Mediterranean, the Po to the Adriatic, the Danube to the Black Sea and the Elbe and the Rhine into the North Sea. The vastness of the European Soviets is drained by major rivers which mostly flow across low lying land: the Volga debouching into the Caspian Sea, the Dvina into the Baltic at the Gulf of Riga and a series of great northward-flowing rivers reaching the Arctic Sea.

Each of these major basins has elements of the flora and fauna to a large extent characteristic of it. Thus many species of fishes which are not found elsewhere in Europe occur in the Danube basin, and the rivers of the north and east which are frozen throughout the winter months are inhabited by animals capable of surviving the intense cold by hibernation or some other means. This section of the book attempts to describe some of the major faunal elements associated with rivers, but the visitor to or student of any one major system will find many animals unknown outside that particular area.

The course of any of the big European rivers may be divided into a series of reaches, each characterised by the slope, the speed of flow, the temperature and the amounts of oxygen and silt carried by the water. In the high mountains, where many rivers rise, we find the Torrent Reach. Here the tiny springs originate and bubble down the hillside, often flowing as rapids over bare rocks and boulders or forming waterfalls as they tumble over the edges of cliffs. The water is icy cold, but well oxygenated, and never seriously polluted. The amount of silt carried is very small but minerals may be taken up in solution from the rocks. Little life is sustained in these highest streams for the flow of water is too rapid for plants to take root although a few specially adapted animals including some species of Caddis Fly may be found there. Below the Torrent Reach is the Trout Reach which may reach up to 2000 m. in the mountains but equally forms the headwaters of lowland streams. Generally the slope of the hillsides is still steep here and the water is fast flowing with rapids and small pools and is well-oxygenated, but the erosive force of the flow has broken down some of the rocks to small boulders and pebbles. These accumulate in pools and areas of slack water and the trout which gives its name to this part of the system breed here. In some streams the Trout Reach is extensive; in others as the stream begins to widen and its flow to slacken as the slope becomes less steep it passes rapidly into what is know in mainland Europe as the Grayling Reach and in Britain, where the Grayling is a rather local species, as the Minnow Reach. Here the water rarely reaches more than 10° centigrade in the summertime but is still well oxygenated and swift-flowing with little silt. Salmon breed in this part of the river and a number of other fish species occur. As the stream becomes a river, so the speed of flow slackens further; silt increases and its oxygen carrying capacity decreases, but many fishes inhabit the next Barbel Reach which is the upper of the lowland reaches. Below this again the Bream Reach is the richest in life, the many species of fish characterised by their heavy thick-bodied slow-moving way of life, many of them feeding on the abundant plants of the reach. The Bream Region passes into the estuary, where salt water invades the river at low tide. Some organisms can stand the brackish conditions but many are excluded by them. There is, however, considerable overlap between the reaches and animals characteristic of one are often found in the areas below or above. Few rivers in Europe are unaffected by man. He has frequently embanked the lower reaches to prevent flooding, as with the Thames. Water is removed, often in great quantities, for irrigation, industrial or domestic use, and many upland river valleys are dammed to provide reservoirs which may change the whole character of the flow below the dam. Rivers are often used to remove domestic and industrial waste, and may be heavily polluted sometimes to an extent that no living thing can survive in them. The causes of pollution are now well understood; curing the situation may be a very long term job, requiring in many instances, international co-operation and goodwill. In most cases man's intervention has reduced the aquatic flora and fauna, but where large lakes have been formed in the upper areas these often remain ice free so that the winter bird population has increased in some parts of Central Europe. In general, there is much overlap between the flora and fauna of the slower-flowing reaches and that of ponds and lakes. It is only in the fast-moving waters of mountain streams that we find many special adaptations among both plants and animals to the environment of running water.

WORMS

HAIRWORM —
Gordius sp.

In quiet backwaters, or on the edge of ponds, Hairworms may be found in very great numbers tangled together in a Gordian knot from which their Latin name is derived. Brown, grey or black in colour, and scarcely thicker than a strand of hair although up to 200 mm. long, they have given rise to many superstitions to the effect that they are horsehairs fallen into the water and come to life. In fact these animals are parasitic for at least part of their lives. For the tiny larvae which hatch from the strings of eggs laid in the water by the female Hairworms attack the immature stages of water insects, boring into them and living as internal parasites. Later they infect land living insects, but return as adults to the water, where they no longer need to feed and after a short reproductive life die.

MEDICINAL LEECH —
Hirudo medicinalis

In many respects the Medicinal Leech is similar to the Horse Leech (see page 79), although its colour is usually olive green on the back surface with brown stripes and black spots. It can contract its body to about 20 mm. in length but when extended may measure 150 mm. It is the only European Leech with jaws powerful enough to pierce the skin of large vertebrates on whose blood it feeds, so it is usually found in backwaters and quiet places where cattle or horses come to drink. In Britain it is now rare, occurring in widely scattered localities from western Scotland to the New Forest, its one-time abundance having been diminished by collecting for medicinal purposes. From Roman times onwards Leeches were used to draw off the 'bad blood' from sick people and although this may occasionally have had a good effect in lowering high blood pressure, more usually it resulted in weakening the patient. For when a Leech bites it injects an anti-coagulant so that there may be a considerable loss of blood from a single attack. Furthermore, diseases were sometimes passed from one sick person to another by the use of infected Leeches. Even so, their use continued well into the last century, and when British sources of supply were exhausted, they were imported in very large numbers from Europe where they were farmed in some areas although this was never practised in Britain. The young of Medicinal Leeches hatch from egg cocoons laid in the river bank, near the edge of the water. When immature they feed on the blood of amphibians and small fishes, graduating to large vertebrates when adult.

MOLLUSCS

PFEIFFER'S AMBER SNAIL —
Succinea pfeifferi

The Amber Snails are small waterside creatures with thin, yellowish coloured shells. In this species, which measures about 10 mm. high and 6 mm. wide, there are three obvious whorls, the last very much larger than the others. Identification may not always be easy since in detail the species is very variable. It is found throughout Britain, although it is scarcer in the north, and is widespread over Europe, parts of temperate Asia and North Africa. It lives in and by damp places but is often to be found out of the water on vegetation or on the back of riverside willow trees, sometimes in quite large numbers. It hibernates through the Winter in a protected place near to the water and will hide away also in periods of extreme summer heat.

Limnea palustris

This snail, which occurs through most of Britain although it is rarer in the north and west, is also known through most of Europe, temperate Asia and North America. It is an inhabitant of marshy areas, near to ponds, ditches or rivers and is often found out of the water. There is considerable variation in the shell colour and a number of varieties have been named according to these differences. When full grown the shell measures up to 24 mm. high and 11 mm. broad. The eggs are attached in gelatinous capsules to water plants, which are its main food. In times of drought *Limnea palustris* protects itself with a layer of hardened slime spread over the shell opening.

Unio tumidus

Unio tumidus is one of a number of large bivalve shells which occur in slow-flowing rivers over much of Britain and Western Europe. It can be distinguished from the Swan Mussel (see page 81) by its heavier shell and the notched or toothed structure of the hinge line between the two halves of the shell. In general it grows to a length of about 80 mm. and a height of about 40 mm. and its greatest thickness is slightly less than its height. The life history of *Unio tumidus* is similar to that of the Swan Mussel with the development of glochidia (larvae) which are for a short time parasitic on fish.

A close relative is the Painter's Mussel, *Unio pictorum*, a creature with a very variable shell shape dependent to some extent on its environment, which may include stagnant waters. This shell gets its name from the fact that artists used to keep their paints inside them.

PEARL MUSSEL —
Margaritifera margaritifera

The Pearl Mussel is found in fast-flowing rivers of north and west Britain and throughout much of Europe other than the Mediterranean area, and across Asia to Japan. It also occurs over much of North America as far south as California. The shell may be up to 110 mm. long with a height of half this, but is thinner from side to side than the previous species. It prefers to live in a metre or so of water in an area where the river bed is of sand or fine gravel, and is found particularly in the sheltered area behind big boulders in the river bed. It is not normally found in hard water areas but the soft water of streams flowing over ancient granites and basalts suits it well. In common with most bivalves it can swamp any irritation between the shell and the body with nacre to make a pearl. Sometimes these are quite large and valuable. In Roman times British pearls were

famous but the industry is now nearly dead. The River Pearl Shell has been introduced to some European countries in order to cultivate pearls.

SPIDERS

LONG-JAWED SPIDER —
Tetragnatha extensa

In damp places by streams or ponds the fragile web of the Long-jawed spider may often be seen in Britain and western Europe. This animal may be recognised by its slender body which exceeds 10 mm. in length and the first and second pair of legs which are much longer than this, and which, held out in front of the creature, make it look like a small twig caught in the web. Both sexes have long secateur-like jaws, but those of the male are larger and are provided with hooks which enable him, during mating, to hold the mouthparts of his partner open. This keeps them out of harms way until he can make an escape.

CRUSTACEANS

FRESHWATER SHRIMP —
Gammarus pulex

Several species of Freshwater Shrimp are widespread in Britain and Europe, occurring in streams and shallow pools of many kinds. Typically they are flattened from side to side and normally swim, often with surprising agility, on their sides. The male *Gammarus pulex* is about 20 mm. long, and often carries a female curved under his body. She in turn holds the eggs and for a few days after they have hatched, she carries the young under her thorax. These resemble the adults and have no larval life. Freshwater Shrimps seem never to be still; even when at rest the long antennae on the head may be in movement testing the water for food or foes and the thoracic legs are constantly moving to supply the gills which lie at their base with a supply of fresh oxygenated water. The food of these animals is mainly decaying organic material, but they quite often attack small living animals and have been observed feeding on their own kind. In Winter *Gammarus* hides under stones or buries itself in the mud for protection while the cold weather lasts.

CRAYFISH —
Astacus astacus

The Crayfish is a giant compared to most other British and European freshwater crustaceans, but is small compared with its close marine relatives, the lobsters, which it resembles in many details of its structure. A large Crayfish may measure up to 200 mm. in total length and is most likely to be found in fast-flowing, well oxygenated water in limestone country. Here the calcium salts necessary for its heavy shell may be obtained easily. In many places, Crayfish have disappeared from their former haunts, for they are susceptible to pollution and are easily killed by impurities in the water. Crayfish are mainly active at night when they scavenge dead animal food and hunt small creatures such as worms and water snails. They are themselves hunted by a variety of flesheaters including man, for over much of Europe they are prized as a gastronomic delicacy. In Autumn mating takes place and the female then lays up to a hundred or more eggs which are attached to small abdominal legs called swimmerets. Unlike marine lobsters, Crayfish do not pass through a free larval stage, but the tiny hatchlings, replicas of their parents, cling to their mother for a short time before beginning to lead an independent life.

CHINESE MITTEN-CRAB —
Eriocheir sinensis

The Mitten-crab was originally a native of China, but was accidentally introduced into Europe in the early years of this century. It grows to a maximum width of about 90 mm. and may be recognised by the claws, which are densely hairy, especially in the males. It is found in shallow water in canals and rivers where it makes burrows in the banks. It migrates to the estuaries for breeding, for the larvae must develop in water which is brackish. They migrate back to freshwater when two years old. The

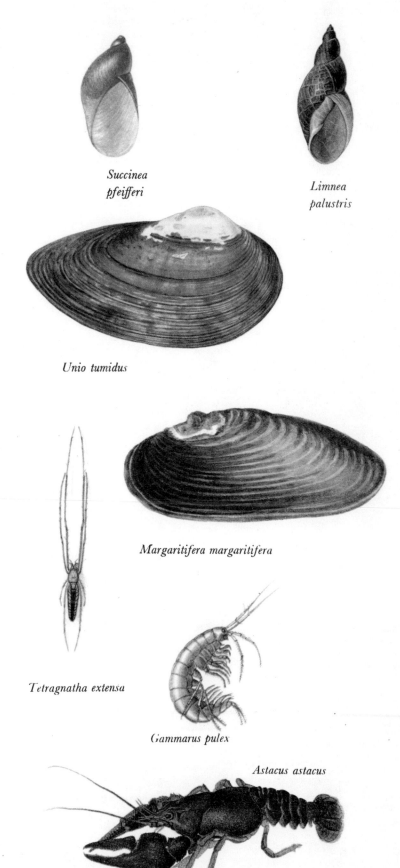

Succinea pfeifferi

Limnea palustris

Unio tumidus

Margaritifera margaritifera

Tetragnatha extensa

Gammarus pulex

Astacus astacus

Eriocheir sinensis

Mitten-crab was once recorded in Britain but fortunately has not established itself here. In the rivers of North West Europe and Scandinavia where it has done so it has become a pest, undermining river banks and damaging fish nets and caught fish.

INSECTS

MAYFLY —
Polymitarcis virgo

In August and September this Mayfly is in some years amazingly abundant over much of Europe, fluttering to lighted windows and round street lamps. It is easily recognised by the pure white of its wings. It does not occur in Britain, apparently needing the flow of great rivers to sustain it in its larval life when it feeds on minute aquatic animals. The life history is similar to that of many British species (see page 82) and after a short adult life it mates and dies.

DEMOISELLE AGRION —
Calopteryx virgo

The emerald green female of this beautiful Demoiselle Fly may be mistaken for that of *Calopteryx splendens* (see page 84), which it resembles closely, although the males are clearly different. In Britain it is found chiefly in southern counties although it does occur sparsely elsewhere. In mainland Europe it may be seen from France to Russia and eastwards to Japan, although this form does not live in the most southerly parts of the continent. It prefers rivers and canals

with a fair flow of water, although it sometimes lives by ponds or lakes. It has a longer season than many other dragonflies and may be seen as early as May or as late as September, but with its weak and fluttering flight it is never far from water. After mating, the female lays her eggs inserted into the leaves of water plants and the nymphs mature in the water. Like their parents they are considerable carnivores.

COMMON COENAGRION —
Coenagrion puella

There are several kinds of blue damsel flies which are common over much of Britain and Europe and close examination may be necessary to be certain of the species. The male Common Coenagrion illustrated here has a wingspan of 45 mm. and a body length of 35 mm. The female which is about the same size is differently coloured with a green and black abdomen. They are on the wing from May until August or even September and although they spend part of their early adult life away from the water, it is here that they are most often seen, fluttering along the edge of ditches, canals and slow flowing streams and ponds, frequently resting on floating or waterside plants. After pairing the male remains with his mate as she lays eggs in the tissues of water plants. She sometimes climbs below the surface for this and is helped back by the firm grasp which he has on her neck. The nymphs are aquatic and take one year to mature.

EMPEROR DRAGONFLY —
Anax imperator

This large dragonfly, which has a wingspan of 100 mm. and a body length of 75 mm., is to be seen flying from early Summer until well into August or even September. The male, figured here, is largely blue in colour with a black stripe down the abdomen; the female is greener. In Britain it is a southern species; in Europe it occurs almost everywhere except the far north of the continent. Southwards it extends into Africa and eastwards to North India. It is a very powerful flier and is often seen well away from water, and is also one of the species of dragonfly which is recorded among swarms of migrating insects. It feeds on many flying creatures, including smaller dragonflies and butterflies which it catches and eats in flight. Although not remaining attached to the female after mating, the male frequently stays in the vicinity as she lays her eggs. Normally the nymph takes two seasons to mature but may do so within a single year.

SCORPION-FLY —
Panorpa communis

Although only about 15 mm. long the male Scorpion-fly such as is illustrated here is a fearsome looking creature, for it carries the enlarged end of the abdomen turned up over its back like the sting of a scorpion and its head is drawn out into a long beak which gives it a fierce, predatory look. In fact it is entirely harmless to man, feeding on other insects and the tail 'sting' is really part of the mating apparatus and is not present in the female. It is to be seen flying during the summer months on the edges of damp woods or along hedgerows. The eggs are laid in the ground and hatch into greyish brown caterpillar-like insects. These feed mainly on other inhabitants of the soil in wet woodlands where there are plenty of rotting tree stumps.

CADDIS FLY —
Limnophilus rhombicus

The adult Caddis Fly, shown in the illustration, although widespread in Britain and Europe, is probably less well known than the larval stage, for it rests by day and becomes active at night, sometimes flying to lighted windows. The larva, like those of all Caddis Flies, is aquatic, and may be seen in many slow-flowing streams where it may be recognised by the criss-cross arrangement of the scraps of plant with which it builds its protective case. Pupation takes place in the case and when the adult insect emerges in late Summer, it bites its way out, swims to the surface and is able to fly almost immediately.

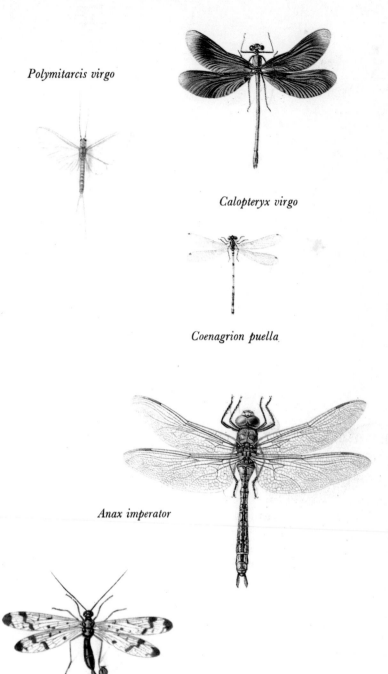

Polymitarcis virgo

Calopteryx virgo

Coenagrion puella

Anax imperator

Panorpa communis

Limnophilus rhombicus

MUSK BEETLE —
Aromia moschata

The metallic colours of the Musk Beetle, which may be blue, coppery, or as in the illustration, green, camouflage it perfectly on the branches of willow trees where it is most often found. The eggs are laid on old willows, which is where the larvae develop and these may even destroy the tree with their tunnels from which the 35 mm. long beetles eventually emerge. Although difficult to see on its host plant, it may sometimes be observed on riverside flowers and at all times be detected by the powerful musky smell which it produces. In Britain this species is less common than in the past, but although rare occurs in widely separated localities. In Europe it is found where ever the habitat is suitable throughout the continent.

Lamia textor

This beetle, less colourful than the previous species and smaller, measuring between 20 and 30 mm. long, is also to be found on riverside willows and related species, where the larvae feed by boring into the wood, although it is rarely deep enough to do serious damage to the trees. The adult rests during the day time and becomes active at dusk, but it is not likely to be met with in Britain, where it occurs sparsely in a few localities only. In mainland Europe it is widespread throughout the continent.

Saperda populnea

This beetle, which measures up to 15 mm. long, is associated with various species of waterside trees, including alders, aspens, poplars and also birch. The adults are to be found in the early part of the Summer, which is when the females lay their eggs on the growing shoots of healthy young trees. As these are uncreviced and smooth she bites a small depression in the bark to contain the egg and the larva feeds at first on the gall tissue which grows round the wound. It then eats its way down inside the young shoot for a distance of about 5 cm. before pupation. Where large populations of this beetle occur they can cause

a great deal of damage to young trees, for each larva destroys the growth of the year in the twig in which it is living. In Britain it is found in only a few localities in the south of the country. In Europe it is widespread, and is found eastwards into Siberia. It also occurs in North America from Canada to New York in the east, and Arizona in the west of the continent.

Oberea oculata

The draining of wetland habitats is probably chiefly responsible for the disappearance of *Oberea oculata* from Great Britain where it is thought to survive only in the fens of Cambridgeshire. In Europe, however, it is widespread and found as far north as Scandinavia. It is a species of the middle and late summer months, when the adults emerge from the willows and sallows in which they have passed their larval life. Although they measure up to 20 mm. long and are brightly coloured and obvious, they often occur in considerable numbers on the branches of trees and have the curious habit of producing creaking noises by rubbing the head against part of the thorax. The eggs are laid singly on twigs and the larvae excavate tunnels up to 30 cm. long. This takes place during the course of the year while the larvae are developing.

Clyta quadripunctata

This curious beetle, about 10 mm. in length, may sometimes be found on trees by streams draining forests and heathlands, where the larvae spend their lives in the nests of ants. The female protects the egg in a case of clay pellets, which the larva enlarges as it grows. Although the ants would probably attack it if they could, the larva retreats into its case and is rarely caught. It feeds on the eggs of the ants and any other food it can scavenge. It is widespread on the Continent but is uncommon in Britain.

Melasoma populi

Easily recognised by its red colour, the 10 mm. long *Melasoma* may sometimes be found in large numbers on the branches of young poplars or willows. Both the adults and the larvae feed on the leaves and may do considerable damage to the trees in this way. The larva defends itself with the production of an extremely foul smelling secretion, which is distasteful to birds and other predators. The species occurs widely in Europe and non-tropical Asia and is found as far east as Japan; in Britain it is most likely to be seen in the southern counties.

Agelastica alni

This little beetle, no more than 5 to 6 mm. in length, sometimes occurs in such very large numbers on alders as to cover the branches and twigs completely. The adults, which may be seen first in the early Summer, feed on the leaves to some extent but do not survive for long. The females, recognisable by their abdomens swollen with eggs, lay these in batches of about fifty on the undersides of alder leaves, each female producing about 900 eggs in all. The larvae soon hatch and feed through the early Summer, then descend to the ground to pupate, emerging as adults in August. These return to the alders in which they remain until forced to the ground by the onset of cold weather. They hibernate in the protection of fallen leaves and return to the new foliage next Spring. Although generally abundant in Europe, including most of Scandinavia, this species is very rare in Britain.

PUSS MOTH —
Cerura vinula

Although the adult Puss Moth measures about 70 mm. in wingspan, it is difficult to see as it rests on the bark of trees during the early summer months, for it is perfectly camouflaged by the zig-zag lines

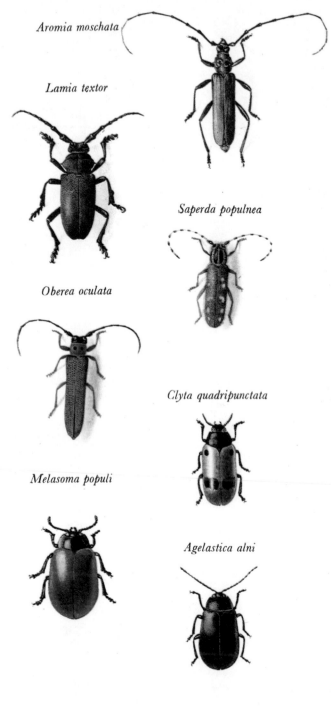

Aromia moschata

Lamia textor

Saperda populnea

Oberea oculata

Clyta quadripunctata

Melasoma populi

Agelastica alni

Phalera bucephala

Cerura vinula

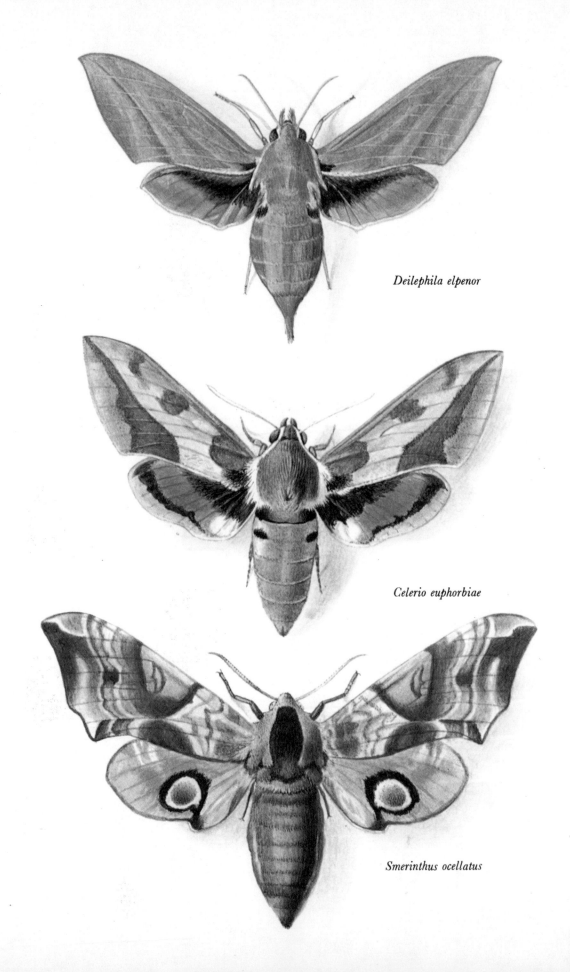

Deilephila elpenor

Celerio euphorbiae

Smerinthus ocellatus

on its wings. The caterpillars which hatch from its eggs and are to be found from midsummer onwards, are also camouflaged in shades of green and purplish-brown, but have a further defence should they be detected by any of the small birds which are their chief enemies. From the forked appendage on the rear end they shoot out bright red filaments which surprise and terrify the attackers, many of which refuse to approach the caterpillar again. The full-grown larva pupates in early Autumn in a cocoon protected by bark splinters and hidden in a crack in a tree trunk. The Puss Moth may be found throughout much of Europe including Britain and non-tropical Asia where its numbers occasionally cause damage in plantations of young trees.

BUFF-TIP —
Phalera bucephala

Defoliated branches of willows, poplars, birches or limes may be the work of the caterpillars of the Buff-tip Moth. These, which are found throughout Britain and Europe, may be recognised by the yellow stripes on their hairy, dark brown bodies. While they are small, the larvae feed together, but as they grow they spread out over the tree. When fully grown they pupate in the ground overwintering there and emerging in May and June. The moth has a wingspan of about 50 mm. but when at rest is one of the most difficult to see, for it wraps its wings over its body so that it is rounded like a piece of twig. The buff

patches on the outer end of each hind wing are exposed to look like the broken end of a rotting twig as does the light coloured head. Sitting openly on the branch of a tree the eyes of predators and humans pass over this wonderfully camouflaged creature which escapes detection more often than not.

ELEPHANT HAWK MOTH —
Deilephila elpenor

Where rosebay willow herb grows in any quantity, the Elephant Hawk Moth is to be found, for it is here that the female lays her egg batches in early Summer. It is from the caterpillars, which grow to a length of about 80 mm., that the species gets its name for in them the head and thorax can be retracted and then thrust out, looking like the trunk of an elephant. When the 'trunk' is withdrawn, prominent eye spots on the sides make it look a most fearsome animal. The adult may be seen throughout the summer months flying at night and feeding on nectar from long-tubed flowers. Perhaps because of the abundance of the larval food plant in urban waste lands, the Elephant Hawk Moth is often found in towns. Apart from this it is widespread in Britain and Europe and non-tropical Asia.

SPURGE HAWK MOTH —
Celerio euphorbiae

This species is found only as a rare immigrant in Bri-

Laothoe populi

tain, but is widespread through Central and Southern Europe where it may be seen from middle to late summer, feeding on long-tubed plants, sea campion being a great favourite. The caterpillar, which is brilliantly coloured in red, black and yellow, feeds especially on maritime species of spurge. Pupation which occurs in the soil is very variable in length, occasionally taking two years. More frequently it takes only one year and sometimes a period of only a few weeks before the adult insect emerges, recognisable by its rather short (about 30 mm. long) plump body and its wingspan of about 75 mm.

EYED HAWK MOTH —
Smerinthus ocellatus

The caterpillars of the Eyed Hawk Moth feed on willows and poplars, and the handsome adult moth may be found in damp woodland areas throughout Europe where these species occur. Although the wingspan is about 85 mm. the moth is rarely seen when at rest, for the forewings match perfectly in colour and pattern the bark on which it is sitting. If detected by a small predator, such as a bird, the moth's reaction is to draw the forewings aside, revealing the brightly coloured hindwings, with their eye markings. These are then vibrated slightly giving the effect of a face moving towards the intruder, the 'snout' being formed of the fat, furry body of the moth. Experiments have shown that this fright effect

is very considerable and in some cases may prevent the bird from attacking the Eyed Hawk Moth again.

POPLAR HAWK MOTH —
Laothoe populi

Poplars, including the strongly scented balsam poplar, are the food plant of this species, which is widely distributed in Europe and is the commonest hawk moth species in Britain. After feeding through the summer months, the large green caterpillar, diagonally striped and dotted with yellow, descends to the ground to pupate, remaining securely in the pupa until the early Summer when the adult emerges. This may be seen until the early Autumn, but is rarely observed during the daytime for the camouflaged colouring of the forewings offers near-perfect protection as it rests on the trunks of trees.

ROSY UNDERWING —
Catocala electa

This moth, which is very similar in appearance to the Red Underwing and its cousin *Catocala elocata* (see page 34), is a European and Asiatic species, associated with willow trees, the leaves of which form the food of the larva. Like its close relatives the adult spends its days on the bark of trees, hidden by the camouflaged colouring of its wings, and is active only at night. Although on the wing in Europe from

Catocala electa

Apatura iris

Argynnis paphia

Limenitis populi

midsummer onwards, the rare records of this moth reaching Britain are almost all for the month of September which is near the end of the season for the species. This is a coincidence for which there is no explanation.

PURPLE EMPEROR —
Apatura iris

This beautiful butterfly gets its name from the purple sheen on the wings of the male; the larger female, with a wingspan of about 60 mm., is brown in colour. It is one of the most spectacular British insects, occurring mainly in woodlands in the southern part of the country, although it is widespread through much of Europe. Its habitat must include at least two sorts of trees, willows or sallows on which the larvae feed, and tall woodland trees where the adults fly. In midsummer, when they are on the wing, Purple Emperors congregate around the tops of the tallest trees which stand out above the general forest canopy. These 'king trees' may be the gathering place for many generations of the butterflies which court and mate at high level. They may occasionally be seen near the ground, where they come down to feed on nectar or to drink the juices oozing from the excrement of a dead animal. After reproduction, the adults die, and the caterpillar, which is a dramatic looking creature with two blue horns on its head overwinters, hidden on a twig of its food plant.

POPLAR ADMIRAL —
Limenitis populi

This handsome butterfly does not occur in Western Europe but may be common in the central and eastern parts of the continent, and is also found across Asia to Japan. The male measures up to 40 mm. in wingspan; the female, which is distinguished by a broad white band on her hind wings, may be larger. Its habitat is open woodlands, where it is on the wing during June and July. The caterpillar, which is chiefly green in colour, feeds on the leaves of poplars or aspens. It overwinters in the larval stage, rolling itself in a leaf as protection against the cold and predators.

Haematopota pluvialis

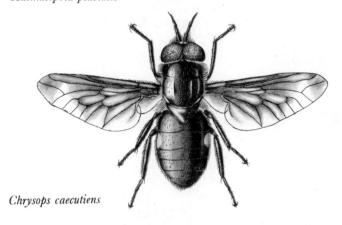

Chrysops caecutiens

SILVER WASHED FRITILLARY —
Argynnis paphia

In Britain this butterfly is found only in Ireland and south of a line from Anglesey to south east Kent, especially in wooded areas. In Europe it is absent from southern Spain and most of Scandinavia and the north, although it is present across Asia to Japan. The butterfly is on the wing in midsummer, when the eggs are laid on tree trunks. The caterpillar hatches in August, and after eating its egg shell starts to feed on the leaves of dog violets. Its growth has not progressed far before it goes into hibernation at the base of its food plant, but in Spring it emerges and completes its development. At this stage it is velvety black with two yellow lines running the length of the back and reddish coloured branched spines tipped with black. The chrysalis is pale brown with a golden sheen and may be active if touched.

Haematopota pluvialis

As with the Horse Fly (page 89) to which it is closely

131

Stratiomys chamaeleon

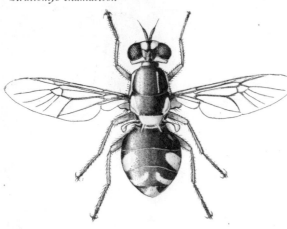

Oxycera pulchella

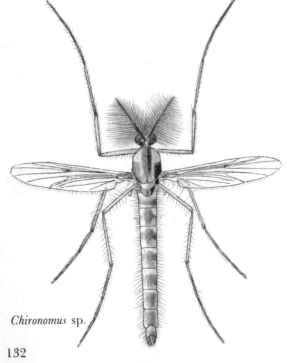

Chironomus sp.

related, it is the female Cleg which needs a meal of blood before she can lay her eggs. The male is an inoffensive creature feeding on nectar and is rarely noticed. His mate, however, makes a silent approach to her victim and using her stout proboscis inflicts a painful bite. The larvae are aquatic and feed on worms and grubs.

THUNDER FLY —
Chrysops caecutiens

This is another species of biting fly not found in Britain but widely distributed over much of Europe and like the Cleg apparently most active in still, sultry weather during the midsummer months. Thunder Flies are slightly more noisy in their approach than are Clegs, but often manage, even so, to take their necessary meal before egg laying. The illustration shows a male; the female is more brightly coloured and has beautiful irridiscent eyes.

SOLDIER FLY —
Stratiomys chamaeleon

Soldier flies are non-aggressive animals and probably get their name from their bright colours reminscent in some species of military uniforms. Although widely distributed it is only locally abundant, flying during the mid and late summer months. It may be recognised by the apparently flattened abdomen, and the long, angled antennae, which is an unusual feature in flies. The eggs are laid on aquatic plants and the larvae climb below the water surface to hunt small invertebrates. They have a long breathing tube, crowned with a rosette of hairs which acts as a float at the surface of the water and by folding inwards traps air when the creature is submerged.

Oxycera pulchella

This handsomely coloured fly, which measures up to 7 mm. in length, may be seen on sunny days resting on the leaves of bushes close to streams in southern England and over much of Europe. The females lay their eggs in still or slow-moving waters where there is some vegetation and the larvae develop in the mud.

NON-BITING MIDGE —
Chironomus sp.

There are many species of small, fragile bodied waterside fly like the Non-biting Midge illustrated. They are abundant throughout the summer months over all of Britain and mainland Europe and the rest of the world. Non-biting Midges are often found near highly polluted streams for the aquatic larvae contain haemoglobin which helps them to survive in oxygen deficient waters. Their red colour has led them to be known as bloodworms. At all stages of their life Chironomids are an important food source for many kinds of water creatures.

FISHES

BROOK LAMPREY —
Lampetra planeri

Brook Lampreys are unobtrusive small creatures, reaching a maximum length of 160 mm. They live in the upper reaches of rivers in Britain and Western Europe. Although included here with the fishes, they are only distantly related to them and are usually placed in a different classificatory class, the Cyclostomes. In Spring the adult females shed about 1500 eggs into a prepared nest in the stream bed. These hatch into larvae known as ammocetes, which live buried in the stream bed where they filter the minute organisms on which they feed. Their devel-

opment may take up to five years. When they finally metamorphose to the adult state they feed no more but breed and die.

SALMON —
Salmo salar

The Salmon is now much rarer in European rivers running into the North and Arctic seas than it used to be. Decimated by pollution and diseases and prevented from reaching its breeding grounds by dams and barrages built across many major rivers, it has become extinct in many places where it was formerly numerous. Fortunately, its value as a game and food fish is now recognised and the position is being slowly improved by cleaning polluted waters, building ladders by which the fish can ascend rivers made impassable by human works and by restocking from fish farms. The Salmon's life is divided betewen the clear, cool waters of the upper middle reaches of rivers and the sea. Spawning takes place in the winter months, in a nest or reed, hollowed out by the female. She and her mate then try to return to the sea, but frequently do not succeed. While spawning the adults do not feed and become seriously weakened during the breeding season, falling easy prey to predators and disease and often with insufficient strength to battle successfully with the river. The eggs hatch in Spring, but the young depend on their yolk sac for some time after this. Later they feed on small invertebrates until at the age of one to five years they travel down stream to the sea,

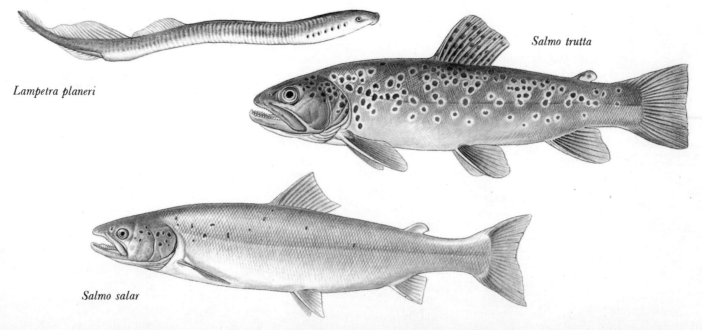

Salmo trutta

Lampetra planeri

Salmo salar

where they feed and grow for another five years before returning to spawn in their natal stream.

TROUT —
Salmo trutta

Several forms of trout have been described from European waters. These include the Sea Trout which is to be found in streams from the White Sea to north Spain. Like the Salmon this is migratory and has suffered much the same fate from pollution and alternation of river beds and courses although occasionally large specimens, up to 1000 mm. in length and weighing 15 kgm., may be caught. Its spawning migrations are later than those of the Salmon and reach further up stream; fewer individuals die after breeding so that most spawn several times. The Lake Trout, *S. t. lacustris* is another migratory form but this lives in large, cool mountain and northern lakes and migrates into the tributary streams to breed in Autumn. The young fish live for up to several years in the stream before migrating to the lakes where they mature in the course of the next few years sometimes becoming at the age of about seven years as large and heavy as Sea Trout. The Brown Trout, *S. t. fario* is the most widespread form, for it is found in suitable streams throughout the continent. Unlike the others it is quite sedentary and does not move far in the course of its life. The size to which the Brown Trout grows is to some extent dictated by its environment. In small streams or where the food is poor, it grows to a length of about 200 mm. and is mature at about three years. In bigger rivers it may take five years to mature but grows to a length of 500 mm. All trout are carnivores, feeding on a wide range of smaller animals, according to their circumstances. In some areas American Rainbow and Cutthroat Trout have been introduced and in a few places these have now become fully naturalised in their new environment.

POWAN —
Coregonus lavaretus

This fish belongs to a complex of species related to salmon and trout which are found in Alpine and northern lakes and some rivers in Europe and known collectively as Whitefish. The Powan occurs in a number of British lakes, where it may be given local names; the Gwyniad in Wales for example or the Schelly in the Lake District. It feeds mainly on tiny crustaceans which it combs from the water with gill rakers. In North West Europe and Northern Russia migratory forms occur which have a life cycle like that of the salmon. These may grow to a size of 700 mm. and weight of 10 kgms. Lake-locked forms are normally much smaller.

CRUCIAN CARP —
Carassius carassius

Originating in Asia, the Crucian Carp is now to be found through most of Europe, including the British Isles; yellowish in colour, it is often mistaken for a goldfish. Its usual habitat is ponds and lakes, where it is capable of surviving a greater degree of pollution and deoxygenation than any other species. It feeds on plants and insect larvae and some planktonic animals. Its shape is variable, according to conditions; where food is abundant it becomes very deep bodied, while in ponds where food is scanty it is stunted and thinner. It may hybridise with Carp (see page 92) but is not used as a food fish.

ROACH —
Rutilus rutilus

Roach are among the most abundant fish in slow-flowing rivers and the edges of lakes where there is plenty of vegetation. They occur throughout Europe except for part of the Mediterranean and the coldest areas of the north. Variable in size, in conditions of overcrowding where there is little food the length may be as little as 100 mm. In good conditions lengths up to 400 mm. have been recorded and the body is deep-shaped. Roach is important as part of the food chain of many valuable fish but is not often used for human food.

IDE —
Leuciscus idus

Although the Ide occurs in some parts of Scandinavia

134

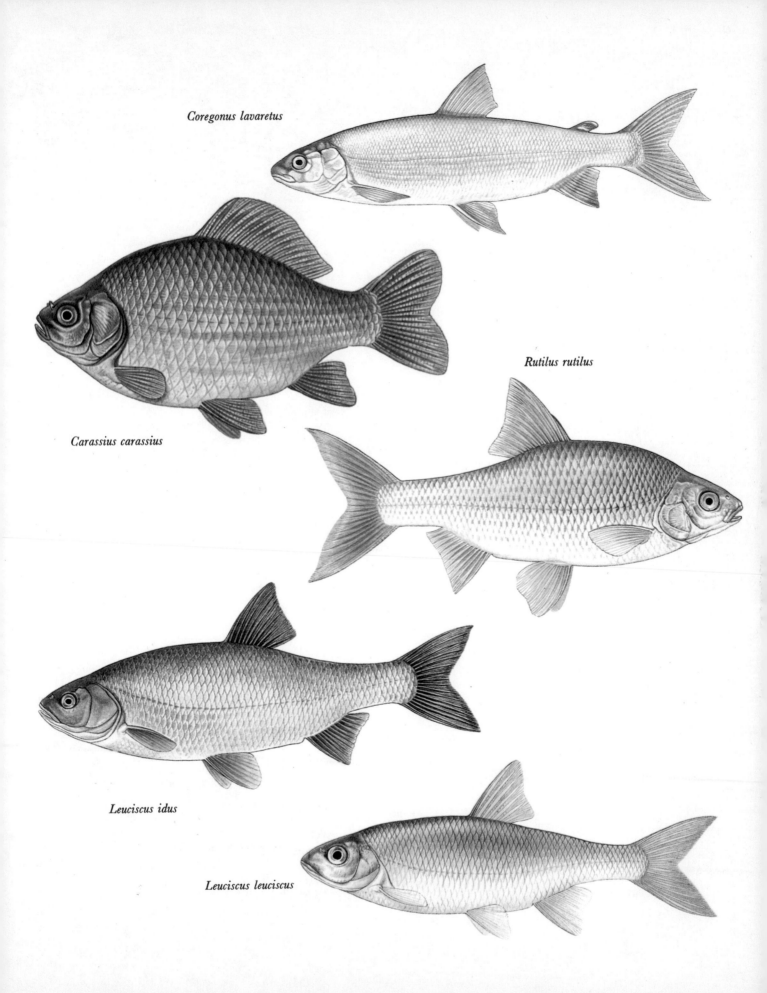

Coregonus lavaretus

Carassius carassius

Rutilus rutilus

Leuciscus idus

Leuciscus leuciscus

it is primarily an Eastern European fish and is found also over much of temperate Asia. Although some are isolated in lakes, it is mainly migratory, living in brackish water and moving into shallow fresh water streams to spawn in the springtime. After spawning the fish move downstream again, often in dense shoals and overwinter in deep water. The Ide may grow to 400 mm. long and specimens up to 600 mm. weighing 4 kgms. have been recorded. It is said to make good eating and in some parts of Eastern Europe is the centre of a considerable fishing industry, although overfishing and pollution have seriously reduced the stocks. In Britain a yellow variety of the Ide, known as the Golden Orfe, is kept as an ornamental pond fish and sometimes escapes into streams although it has not established itself.

DACE —
Leuciscus leuciscus

The Dace is found throughout Europe except for the extreme south and the far north and in Asia as far east as Siberia. Although it prefers cool, fairly fast-flowing streams, where it feeds mainly on small invertebrates of many kinds, it can tolerate other conditions, even the brackish water of estuaries. Spawning is from March to May and during this time the males develop tubercles over the body. The eggs are laid in streams with a sandy or stony bottom and the young reach maturity at three or four years old. Dace rarely grow to more than 200 mm. in length, and since they are extremely bony fish are not much valued for food.

CHUB —
Leuciscus cephalus

In the rapid flow of the Barbel Reach of rivers, Chub may sometimes be found in large numbers. They do not occur in Scotland or the more westerly parts of Britain, but otherwise range through most of Europe south of the Bay of Bothnia. During the summer months Chub live near the surface of the water, often in places shaded by vegetation or overhanging banks. They are extremely wary fish and the least disturbance will drive them into deeper water

where they also spend the Winter. Chub feed on almost any kind of animal food, including fish eggs and fry, which makes them unpopular in trout streams, but they are hunted for their own sake by sporting fishermen who use a variety of baits to attract them including cheese and fruit. A good sized Chub may be 400 mm. long; but some up to 600 mm. occur occasionally.

MINNOW —
Phoxinus phoxinus

The Minnow is the smallest member of the Carp family rarely growing to as much as 120 mm. long. Unlike most of its relatives it is a fish of fast-flowing streams and often forms in small shoals with young trout or salmon of about the same size. It occurs from Britain eastwards across Europe and almost the whole of Northern Asia and is absent only from Norway and the Mediterranean lands. Minnows feed on insects and their larvae, small bottom living invertebrates and fish eggs and fry and are eaten by many kinds of larger fishes and predators such as Kingfishers. Spawning takes place from June to July when the males change colour, becoming reddish below and more intensely coloured above. The females spawn up to 1,000 eggs which are laid in clumps between stones and the little fishes are mature within one year.

RUDD —
Scardinius erythropthalmus

The Rudd, which rarely grows to more than 300 mm. in length and a weight of 400 gm. is to be found through most of Britain and France and southern Scandinavia and from there eastwards across Europe and into Asia. It is a fish which lives amongst dense vegetation in slow-flowing rivers or ponds, although it moves to deeper water during the Winter. It lives in small schools, feeding on the leaves of water plants as well as some insects and other invertebrates. At spawning time in early Summer the female Rudd lays up to 200,000 eggs which are very sticky and adhere to the plants. The young fish hatch in under ten days and are mature at two to three years. Rudd

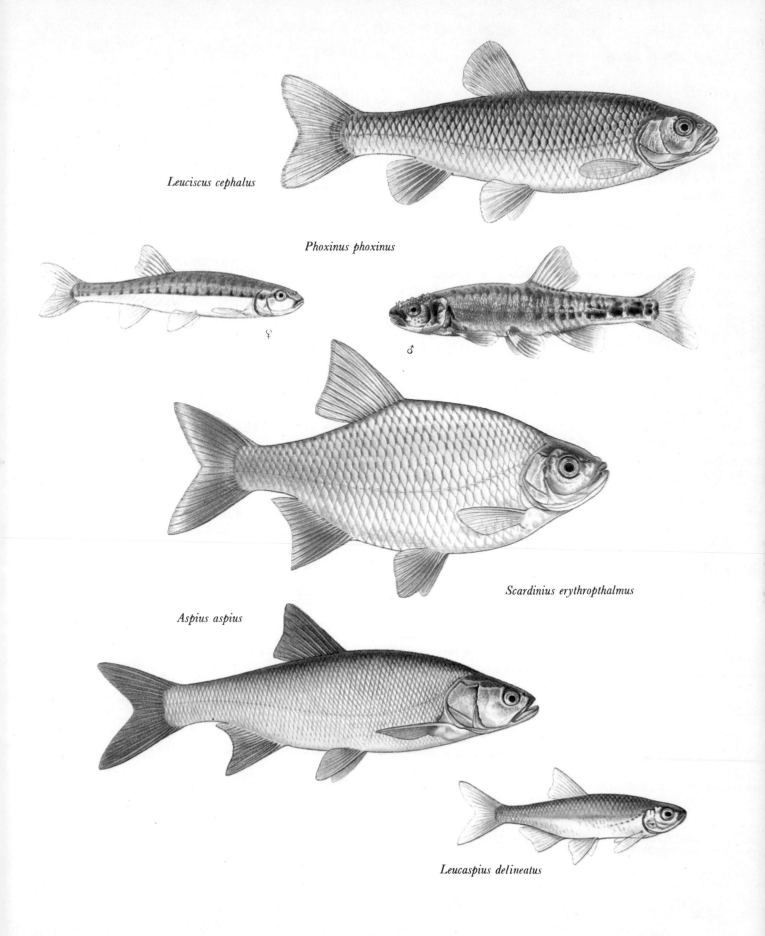

Leuciscus cephalus

Phoxinus phoxinus

♀

♂

Scardinius erythropthalmus

Aspius aspius

Leucaspius delineatus

often swim with other members of the Carp family and frequently spawn with them so that hybrids with Roach, Bream and others are sometimes found. These fish share the characteristics of both parents and may be difficult to identify.

ASP —
Aspius aspius

Although a member of the Carp family, the Asp, unlike most of its relatives, is a solitary predator, feeding on other fishes, frogs and water birds. It occurs in southern Sweden and part of the Low Countries but is essentially a creature of Eastern Europe and parts of Asia. It lives in the middle reaches of lowland rivers, and in the Black Sea and Caspian area is partly migratory. It normally grows to a length of about 600 mm., although a fish 900 mm. long and weighing 9 kgm. has been recorded. Spawning takes place in early Spring in the upper reaches of the river. The young feed at first on animal plankton, but at the age of a few months are already feeding on other young fishes. Asp are highly regarded as sporting and food fish over much of their range.

MODERLIESCHEN —
Leucaspius delineatus

The Moderlieschen, which is to be found from the Low Countries eastwards across Europe, is a small fish, rarely growing to more than 100 mm. long. It lives in shoals in the Bream Reach of rivers and in small ponds, where large numbers of fish seem to appear very suddenly. The German name 'Moderlieschen' which means 'motherless' was given because of this habit of turning up unexpectedly. During spawning, which is in the summertime, the female develops a fold of skin round the vent which acts as an ovipositor and the eggs are laid in spirals round the stems of water plants. They are guarded and aerated by the male. Moderlieschen survive for up to four years, feeding on plankton, algae and small invertebrates.

GUDGEON —
Gobio gobio

The Gudgeon is found in a broad band from southern Scandinavia to north Spain and stretching eastwards across the whole of the Eurasian landmass. It is highly adaptable and occurs in all sorts of waters, although it is always a bottom liver, preferring a stony or sandy bed to the stream or lake. It feeds on small invertebrates and some plant food and rarely grows to a size much exceeding 100 mm. In spite of its small size it is valued as food particularly in France. In early summer the males develop spawning tubercles and mating and egg laying takes place in shallow water.

BARBEL —
Barbus barbus

Barbels may be recognised as bottom living fishes by the four sensory projections on the upper lip which are themselves called barbels. These fish are characteristic of the middle reaches of rivers where the water is clear and the current rapid. They are often to be found in small groups in places where the current is greatest, such as near to weirs, during the daytime. At night they disperse and feed, eating many sorts of invertebrate and small fish and occasionally some plant food. They normally grow to a length of about 500 mm. Although bony, Barbel is highly valued as food over much of its range and many are caught, especially on spawning migration when the fish move upstream to gravelly river beds where the eggs are laid.

BLEAK —
Alburnus alburnus

This fish is to be found over most of England and Wales and much of Europe other than the far north and extreme south. It prefers clear open water in slow-flowing rivers or lakes where there is little vegetation, often living in large shoals near to the surface. Its food includes plankton in the early stages of life, and later small invertebrates are taken. Bleak rarely grow to a length exceeding 180 mm. and while not important as a food for humans, are used in some areas to feed pigs. They are, however, a vital part in the food chain of many predaceous fishes including perch, pike and trout.

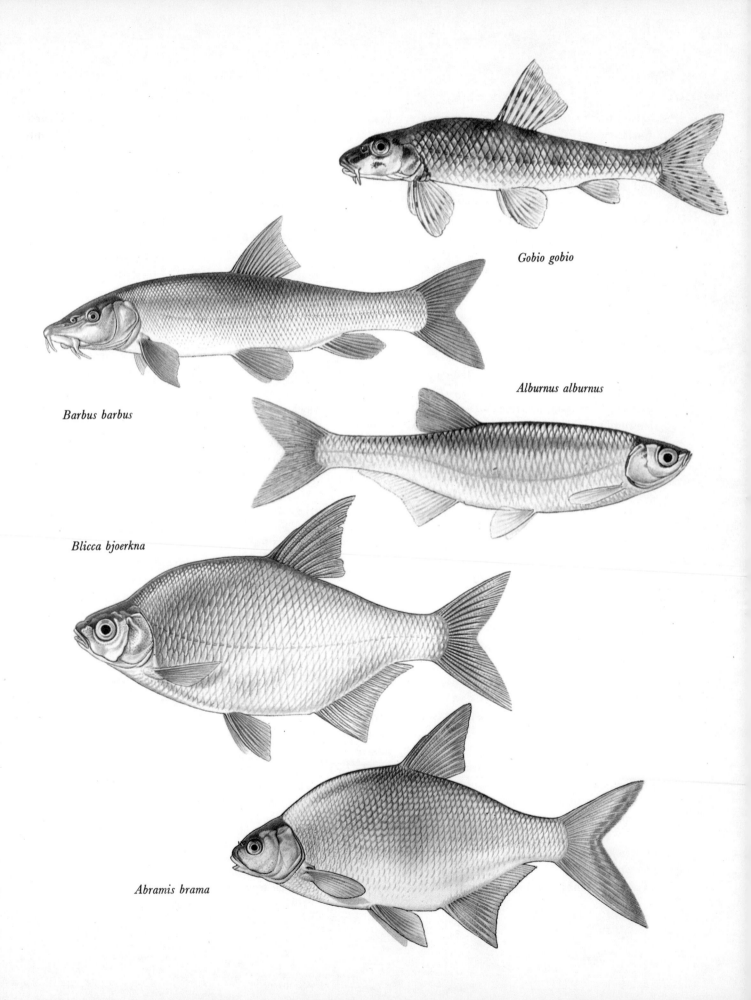

Gobio gobio

Barbus barbus

Alburnus alburnus

Blicca bjoerkna

Abramis brama

SILVER BREAM —
Blicca bjoerkna

This fish is known from some areas of eastern England but is found chiefly in Northern and Central Europe where it prefers shallow stretches of lakes or slow flowing rivers where there is dense vegetation. It competes for food and living space with the Common Bream and various other species. Little valued as an angler's fish, it rarely grows larger than 300 mm. and its food value is decreased by its bonyness.

BREAM —
Abramis brama

This species does not occur in Wales or south west England but is otherwise found in Britain and across Northern and Central Europe, reaching the Mediterranean via the Rhone Valley, and eastwards into Siberia. They live in slow-flowing reaches of rivers and ponds and lakes where the bed is muddy, so they can feed on bloodworms, midge larvae, snails and other bottom living organisms. The mouth of the Bream can be pushed forwards so that the fish can grasp its prey and it habitually turns its body tail-up to feed, leaving dimples in the soft mud in places where it has been sucking in some choice food. In Winter Bream feed less and great numbers of fish may pack together in deep water. Spawning occurs in early Summer, in dense weed found in shallow water. The males, which develop tubercles over the head and front part of the body at this time take up and defend small territories. A large female may lay over a quarter of a million eggs which hatch within a fortnight into plankton feeding young. Growth depends to a large extent on conditions, but Bream are long-lived fish and are among the largest freshwater species found in Europe, having been recorded at a length of 800 mm. and a weight of 9 kg. Over much of the Continent it is highly regarded as a food fish.

ZÄHRTE —
Vimba vimba

Zährte are found in the lower reaches of rivers flowing into the Southern Baltic and the Black and Caspian Seas, as well as the lakes in South East Europe. The river populations migrate during the early Summer to shallow but weedy streams with a stony bed where spawning takes place. Both sexes develop spawning tubercles at this time and the males adopt breeding colours of dark green on the back and sides and orange underneath. Although bony, the Zährte is considered good eating and is widely trapped and netted.

BITTERLING —
Rhodeus amarus

Bitterling are small members of the Carp family, rarely reaching 900 mm. in length, 500—600 mm. being far more usual. It is mainly found in Central and Eastern Europe. It lives mainly in the weedy zones of small lakes and ponds and slow-flowing streams, where it feeds chiefly on plants. Like many small fish, the Bitterling lays only a relatively few eggs, but a high proportion of the young survive because of the care which is taken of them in their early life. In springtime male Bitterlings develop bright irridescent colours and take up a breeding territory but unlike other fishes in which this is spatial, in the Bitterling it consists of a large freshwater mussel (see pages 81 and 122). They will defend this against other fishes and should the mussel move the Bitterling will go with it. Females at this time develop a long egg laying tube or ovipositor and with this lay the eggs, a few at a time, inside the shell on the gills of the mussel. The male sheds milt close by and some of this is drawn in by the mollusc and the eggs are fertilised. The eggs remain protected by the mussel until they hatch two to three weeks later, after which they stay for only a few days before leading an independent life. Their presence appears not to harm the mussel in any way and Bitterlings are among the fish which are used by the freshwater mussels for the development of their young in the glochidia stage (see page 81), so the complement is returned.

STONE LOACH —
Noemacheilus barbatulus

Cool, clear water in rivers and brooks is the place

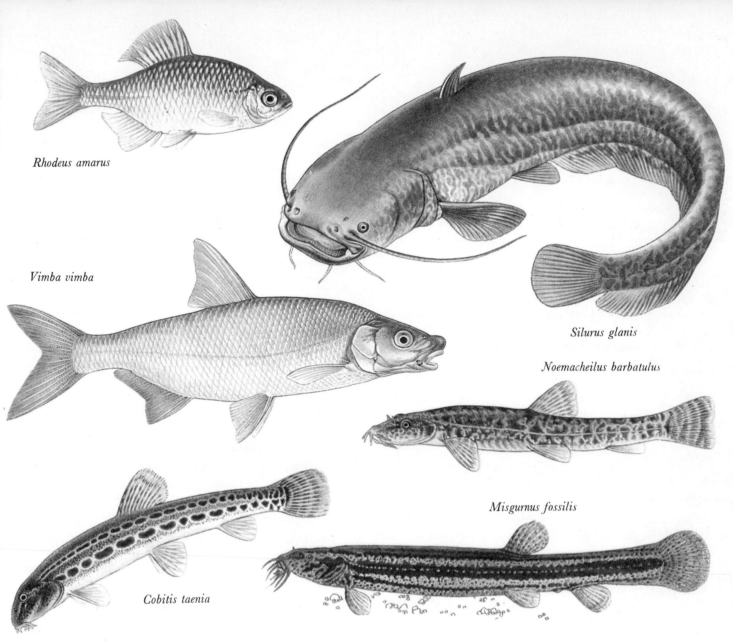

Rhodeus amarus

Vimba vimba

Silurus glanis

Noemacheilus barbatulus

Misgurnus fossilis

Cobitis taenia

where the Stone Loach may be found in Britain apart from Scotland, and throughout Europe except for the far north and extreme south. It is adapted in many ways to spend its days on the stream bed, or even partly buried there. Its eyes, for example, are protected with an extra outer covering against damage from silt. Since it is active and hunts its prey at night, its skin, particularly near the head, is studded with tiny sense organs sensitive to minute changes in pressure so that it is aware of movement in the water whether enemies or prey which are small invertebrates. Spawning takes place in springtime and the female is said in some instances to guard the eggs. Growth is up to 120 mm. in three years

and it rarely grows over 150 mm. long. Although the flesh is said to taste good it is too small to be of much value.

SPINED LOACH —
Cobitis taenia

This fish is found in clear running water or in lakes with a sandy bed where it can bury itself during daylight hours. It occurs in England although it is absent from Wales, Scotland and Ireland. It is also found across most of Europe from southern Scandinavia southwards and through Asia to Japan. Like the other loaches the Spined Loach has a group of sensory barbels on the mouth which help it in finding its prey when it hunts at night, but in this species they are very short compared with those of its relatives. Below each eye is a forked spine, normally hidden, but erected if the fish senses danger.

POND LOACH —
Misgurnus fossilis

The Pond Loach is a fish of Eastern and Central Europe, living in shallow ponds and small lakes and slow-flowing rivers. Feeding on small invertebrates it may grow to a size of 500 mm. but is generally smaller than this. It has many adaptations for life in oxygen deficient waters. One of the most important of these is its ability to breathe atmospheric oxygen and it may sometimes be seen rising to the surface of the water and taking a gasp of air. This is swallowed and some of the oxygen it contains is absorbed in the intestine and passed to the tissues. Should the water dry up the Pond Loach can survive by burying itself in the mud and reducing all its vital functions to a minimum. It is said to be able to survive for a year in this manner. The larvae of Pond Loach have curious external gills, which they lose as they grow, but which help them in the oxygen deficient environment in which they live.

WELS —
Silurus glanis

Catfish are found mainly in the tropics, but the Wels which belongs to this family occurs in Central and Eastern Europe and Central Asia where the summer temperatures are high enough for spawning to be successful. It normally lives in large ponds with a muddy bottom or slow-flowing lower reaches of big rivers, occasionally extending into the brackish waters of estuaries. The male excavates a hollow in the mud in a densely vegetated place near to the bank and here the female lays her eggs which stick to each other and the water weeds to form a sort of nest. This is guarded by the male for a few days until they hatch after which there is no further parental care. The first food of the young Wels is plankton, but they grow very rapidly and soon graduate to feeding on other fishes. At the age of one year they are about 200 mm. long; at the end of four years about 500 mm. and the average at the end of nine to ten years is 1,000 mm. Fish of this size weigh about 10 kgm., although weights much greater than this have been recorded, the largest being 306 kgm. from a fish 5,000 mm. long. Wels feed on anything they can subdue, chiefly other fishes, amphibians and water birds, but there are many stories, mostly unsubstantiated, of dogs and bathers being attacked. The Wels is important commercially in Eastern Europe for the flesh, especially of the young fish, is good to eat and glue and a kind of leather are made from inedible parts of the fish.

CATFISH —
Ameirus nebulosus

Two species of Catfish have been introduced into Europe from North America. The one illustrated is the more widespread and is to be found in many slow-flowing rivers, ponds and swamps in Central Europe and France. Although related to the Wels it is much smaller, rarely growing to much over 400 mm. In midsummer the Catfish pair up and make a nest hollow in a sheltered spot. Here the female lays her eggs, which are guarded by the male who continues to care for the young fish after they have hatched. Catfish are active mainly at night, when they feed on insects and their larvae, molluscs and the eggs and fry of other fish. The flesh is said to be good to eat, but they are not kept in fish farms. Their habit of burying themselves in the mud at the bottom of the pond makes them difficult to catch by normal commercial methods.

EEL —
Anguilla anguilla

Eels are familiar fish in almost all European rivers, where the females may grow to a length of 1,000 mm. and weigh up to 3.5 kgm., although the males are always considerably smaller. The life cycle of the Eel is still known only in outline, for part of its life is spent at sea probably in deep water where none has been caught. It is known that Eels from both American

142

and European rivers reproduce in the Sargasso region, about 4,000 km. from the European coast. The tiny creatures which hatch from the eggs are quite unlike their parents and were at one time thought to be another sort of fish altogether and were given the name *Leptocephalus*. It was subsequently realised that these flat-bodied creatures were the young Eels but they are still referred to as *Leptocephalus* larvae. They make their way northwards on the Gulf Stream, the American stocks which are similar in appearance but which can be distinguished on a vertebra count, diverging and turning eastwards. The rest, however, continue with the North Atlantic Drift until they reach the European coasts. Here they remain in brackish coastal waters for some time, and may be as much as four years old. They have changed by this time into the familiar snake-like shape of the adult Eel. They make their way upstream where they remain and grow for a number of years, feeding mainly at night on all sorts of other living things. In some areas they are regarded as a pest because of their predaceous habits, but in others they are valued as food fishes. Eventually in Autumn, they change again; the colour goes from golden to silver, the eyes become bigger, the jaw muscles and the digestive system waste away and they are ready for the journey down to

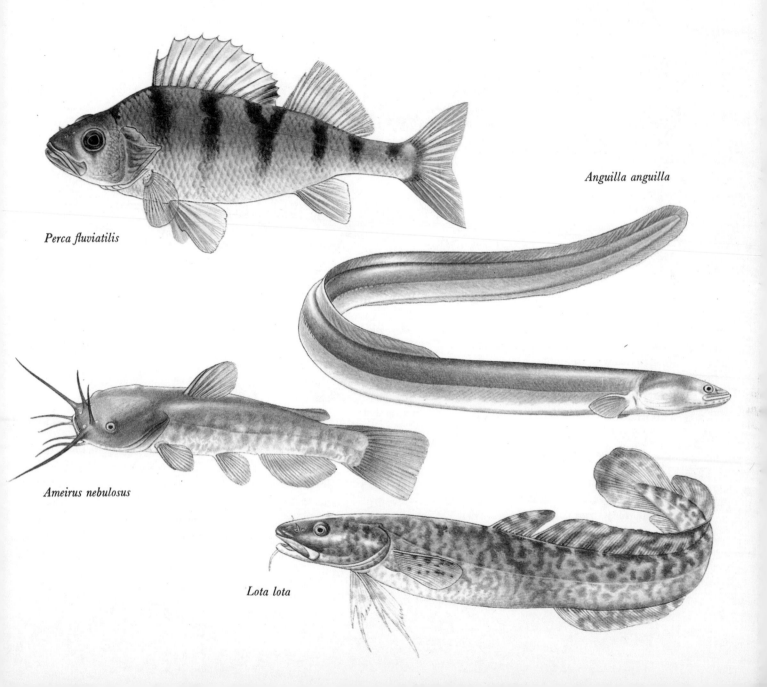

Anguilla anguilla

Perca fluviatilis

Ameirus nebulosus

Lota lota

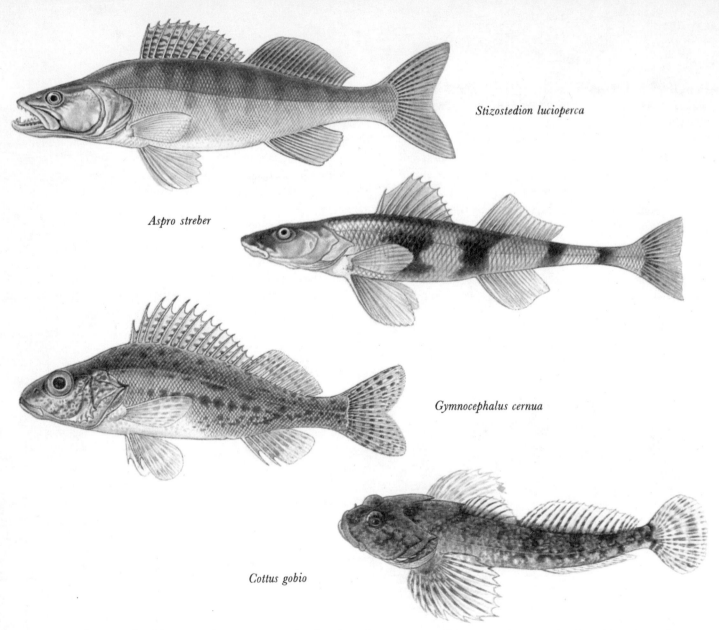

Stizostedion lucioperca

Aspro streber

Gymnocephalus cernua

Cottus gobio

the sea. At this stage they are very fat, for they do not feed again and must travel fast against the main currents to reach the Sargasso Sea by next Spring, where having mated and spawned, they die.

BURBOT —
Lota lota

Burbot are members of the Codfish family but unlike their many marine relatives are found in lakes and cool rivers with a gentle flow and only occasionally in brackish waters. They occur throughout Canada and the northern part of the United States and across the cold and temperate parts of the Eurasian land-mass. They are absent only from Southern Europe, parts of Scandinavia and most of Britain where they are, however, found in a few rivers in eastern England. Their need for cool water is expressed in their inactivity in Summer when they retreat to deeper water and feed less than in the Winter when they spawn. Their food when young includes many small invertebrates and as they grow, other fishes and their eggs. Burbot of a length of up to 1,000 mm. and a weight of 30 kgm. have been recorded from Siberia; but these fishes are estimated to have been about 20 years old. In Europe fish up to six years old may measure 400 mm. and weigh 500 gm.

PERCH —
Perca fluviatilis

Perch are to be found in rivers and ponds through most of Europe except for the extreme north and the far south of the Continent. It also occurs through most of temperate Asia and in part of eastern Canada and the United States. Its preferred habitat is well oxygenated waters in lowland areas, although it also occurs in brackish waters and in some upland lakes. Perch are non-migratory fish and usually live in shoals, although the older individuals may be solitary. They feed on many sorts of invertebrates and, as they grow, on other fishes. When young, they may themselves be the prey of larger predators. Spawning takes place in Spring, usually where there is plenty of vegetation round which the female strings her eggs. These are well protected with a mucous layer and they hatch in two to three weeks. The rate of growth varies with the food supply and in many places the water is populated with large numbers of fish measuring only about 70 mm. although they may be over three years old. Where there is less overstocking, Perch may grow much larger, a good size being 250 mm. at about ten years of age.

PIKEPERCH —
Stizostedion lucioperca

This is a fish of Northern and Central Europe and Western Asia, where it lives in large, well oxygenated warm waters of lakes and the lower reaches of rivers. Spawning occurs in early Summer and both parent fish guard the eggs which hatch in about one week. In their early months they live in shoals, feeding on small invertebrates, but later they become more solitary and feed on other fishes. They are regarded as a valuable game and food fish and have been introduced to waters beyond their normal range, including rivers in eastern England.

STREBER —
Aspro streber

The Streber is a fish found only in the Danube basin and parts of northern Greece, where it leads an inactive life on the bottom of rapidly flowing streams. It is well camouflaged between the stones during the day, and becomes active at night, hunting for worms and other small invertebrates. It grows to a length of only 180—200 mm. and is of no importance for food or as a game fish.

RUFFE or POPE —
Gymnocephalus cernua

Ruffes are found over much of Northern Europe and Asia in slow-flowing streams where there is deep, clear water or in lakes or reservoirs. Here they lie on the bottom at night but are active during the daytime, feeding on insects, crustaceans and the eggs and fry of other fish. In spite of their flesh-eating habits Ruffe rarely exceed 150 mm. in length in five or six years of growth. Spawning takes place socially in shallow water in springtime, after which the adults move further upstream, returning to the lower reaches in the Autumn.

MILLER'S THUMB or BULLHEAD —
Cottus gobio

This little fish is to be found living on the stony bottom of the upper reaches of streams or in clear lakes over much of Europe other than the far north and Scandinavia and the countries of the extreme south. During the daytime it is inactive, but it hunts at night for small insect larvae and the eggs and fry of other fish, including those of trout which makes it unpopular with fishermen. It grows to a length of 180 mm. Spawning occurs in the early part of the year when, stimulated by the courtship display of the male, the female lays her clutch of eggs in a sheltered place where they are guarded by the male until they hatch several weeks later.

AMPHIBIANS

MIDWIFE TOAD —
Alytes obstetricans

The Midwife Toad is an animal found in South West

Europe, from Spain to the Black Forest. It is a small animal measuring little more than 55 mm. at the most. Its habitat is hilly country, where it can find a secure retreat under stones, or in holes which it excavates using its forelimbs and snout. Here it remains during daylight, but it becomes active at dusk, running and jumping as it searches for insects and other invertebrates on which it feeds. The mating season is prolonged and at this time the voice of the male may be heard in a musical call which has caused the species to be known as the Bell Toad. The female may produce several batches of spawn during the course of the season. On each occasion a string of a few (not more than a hundred)

Alytes obstetricans

Rana ridibunda

Emys orbicularis

rather large eggs are produced. These are taken by the male who wraps them round his hind legs and carries them with him on land for two or three weeks until they are ready to hatch. Only then does he go to water to release the little tadpoles which swim away independently.

MARSH FROG —
Rana ridibunda

This is the largest species of European frog and may measure up to 170 mm. long in the southern part of the continent. In general it is a species of Eastern Europe, reaching Germany and Holland in the north and from here it has been introduced into some places in southern England. Its life history is similar to that of the Edible Frog (see page 96), but it prefers weedy slow flowing rivers or lakes and ponds with plenty of vegetation. The male has large vocal sacs and in the breeding season in early Summer is very noisy. The Marsh Frog is a carnivore, eating a wide range of foods, including such small birds and mammals as it can subdue.

REPTILES

EUROPEAN POND TORTOISE —
Emys orbicularis

The European Pond Tortoise is found from North Africa through much of Southern Europe into north Germany and parts of Western Asia. It lives in and by slow flowing muddy rivers and in ponds and pools and may be seen sunning itself. It will plop into the water at the least disturbance. As with many water tortoises the shell is somewhat reduced and the upper and lower halves are not as rigidly joined as in land living forms. Another adaptation to water may be seen in the webbed feet. The size of this animal varies according to the area in which it lives. In the south it may grow to a length of 350 mm. while in the north, where conditions are less suitable for its survival and where the animal will spend longer in hibernation in the mud at the bottom of the streams, specimens 250 mm. long are at maximum growth. European Pond Tortoises are carnivores, feeding on frogs, fishes and invertebrates which they catch in the water. The illustration is of a female, recognisable by the length of her tail; in the male it is longer in relation to the total body length. In early Summer the female makes a shallow pit near the edge of the river in which she lays up to fifteen eggs. The length of incubation varies but it is never less than eight to ten weeks before the little tortoises hatch.

BIRDS

WHITE WAGTAIL —
Motacilla alba

The White Wagtail belongs to a very variable group of birds. The specimen illustrated is the form which is common over the whole continent of Europe during the summer months although it migrates from the northern area during the wintertime. It is occasionally seen in Britain, but a subspecies, *M. a. Yarrelli*, the Pied Wagtail, is the common form there. This differs from the mainland bird in the colour of its back which is black in summer and a dark grey in winter. It is sometimes seen in coastal areas opposite Britain but never far inland. The White Wagtail, which measures about 80 mm. long, is a streamside bird which feeds on small insects mainly found on the ground. Its name is highly suitable for as it walks or runs it flicks its tail and bobs its head in a way unlike that of any other group of birds. Outside the breeding season Wagtails may often be seen in small groups. The Pied Wagtail seems to find well kept lawns a particularly suitable place for food hunting and they are frequently found in large suburban gardens. At night communal roosts may be formed, numbering in some cases many hundreds of birds. A favourite place for this is in old buildings often in the middle of cities.

GREY WAGTAIL —
Motacilla cinerea

The Grey Wagtail is a more brightly coloured bird

Motacilla cinerea

Cinclus cinclus

than its name would suggest, but is distinguished from its cousin, the Yellow Wagtail, by the dull colour of its back. It is a bird of Western Europe found from North Africa and Spain across to Poland in the north and the Caucasus in the south. It is a fairly sedentary bird and migration is on a small scale, although there is some movement of the Caucasus birds to Western Asia in the summer months. In size and general habits it resembles the White Wagtail, although it is less social.

DIPPER —
Cinclus cinclus

The Dipper is a bird of mountain streams throughout Europe, Asia minor and the Urals. About 175 mm. long, it is reminiscent of a large Wren, but its behaviour is entirely different. When perched by the water it bobs constantly, then darts into the water to seize the insects or crustaceans on which it feeds. It has the ability, unique among birds, of walking under water. It does this by heading into the current and spreading its wings so that the flow holds it down. The Dipper's nest is built among stones or crevices near to the water. A favourite place is in the cliff behind a waterfall, so that the young are protected

147

Luscinia svecica

Motacilla alba

Locustella naevia

Luscinia megarhynchos

Sylvia nisoria

from prying eyes by a curtain of spray, through which the parent birds must fly to feed their brood. Special oil glands produce a substance which protects the feathers against constant submersion. The specimen illustrated is a continental bird; the British form has more chestnut on its underparts.

BARRED WARBLER —
Sylvia nisoria

A bird of Eastern Europe, the Barred Warbler is found in moist wooded areas, such as the edge of a forest or in open country where there is plenty of scrub cover. It measures about 150 mm. long and is easily recognised by the close barred pattern of the breast feathers, although this is less pronounced in the female than in the male which is illustrated here. The melodious song is like that of the Garden Warbler, but is given during a song flight, rather than from a perch.

RIVER WARBLER —
Locustella fluviatilis

This retiring little bird, which measures only about 135 mm., is a summer migrant to Eastern Europe, where it lives and rests in dense cover near to rivers in forest areas. If seen, it may be recognised by the warm brown colour of its back, the clearly defined white eye-stripe and the slightly speckled breast. The song of the male is a trilling tinkle of notes, less varied and musical than that of many warblers. It is given from the top of a bush or shrub protected by tree growth above it. The nest is usually built near to the ground, and the young are fed on the same sort of small insects as nourish the adults. At the end of the nesting season the River Warbler returns to Africa where it passes the winter months.

GRASSHOPPER WARBLER —
Locustella naevia

Although it is often found by water, the Grasshopper Warbler may equally live in dry places, provided that dense, scrubby cover is present. Measuring about 125 mm. in length, it is rarely seen even where it is common, for it is extremely secretive and prefers

to hide rather than fly even when danger threatens. It migrates from Africa during the summer months to occupy suitable places in most of Europe from Ireland eastwards, but does not reach the north of Scotland or Scandinavia nor the dry lands bordering the Mediterranean. The presence of this species may be detected by its song, which is a sustained insect-like churring noise given from a low song post, mainly at dusk or in the early morning. Although it may last several minutes, the song is difficult to pinpoint, since the bird moves its head when singing, which increases the ventriloqual effect of the sound. As with other warblers Grasshopper Warblers and their young feed on insects and their larvae.

NIGHTINGALE —
Luscinia megarhynchos

This legendary singer is a dull brown bird measuring some 160 mm. long, which skulks in the deep undergrowth of the damp woodlands in which it lives. It arrives from migration in April, but does not travel further north than central England although it may be found throughout Europe south of this latitude. In spite of its name the song of the male may be heard by day as well as night. The song is wonderfully varied and rich and may sometimes be consciously modified by an individual bird. This leads to areas of particularly fine song where Nightingales immitate an especially accomplished singer whose presence may be felt for several generations. The nest, built of twigs and roots and lined with hair and moss, is always close to the ground but well hidden in thickets and shrubs. Here five or six olive-brown eggs are incubated for thirteen days by the female. Adults and young feed on insects, worms and spiders.

BLUETHROAT —
Luscinia svecica

In spite of some populations in central Spain and western France the Bluethroat is essentially a bird which migrates from Africa to the far north, nesting in Scandinavia, North Russia and East Europe. The bird illustrated is the northern form; those from a more southerly area have a red, rather than a white

Riparia riparia

Alcedo atthis

spot in the centre of the blue bib. Both subspecies may be found in scrub along the edge of rivers or lakes and sometimes in dryer places. Like the Nightingale it stays well hidden, but its musical, high pitched song is an indication of its presence.

SAND MARTIN —
Riparia riparia

Sand Martins are to be found throughout Europe except for the mountainous areas of north Scandinavia. Wherever there are steep sandy banks beside rivers, lakes or gravel pits, they can make their nest tunnels. These little birds which measure only about 125 mm. long, return to Europe from Africa in April and set about refurbishing their nests, digging out tunnels in suitable places to a depth of 1—1,5 m. at the end of which the nest chamber is made. Because suitable places may not be common, Sand Martins may often nest in considerable colonies. Their flight is fast and acrobatic as they catch insects on the wing to feed themselves and the brood of young which hatch from the white eggs. Sand Martins are among the most heavily parasitised of birds since their nest holes remain for years with little change and make excellent shelters for parasites. The parasites infect the birds during their stay in Europe, but remain dormant while the Martins migrate to Africa.

KINGFISHER —
Alcedo atthis

The most brilliantly coloured of European birds, the Kingfisher is resident over most of the Continent in areas where clear streams and ponds are unfrozen during the Winter. It spreads some distance into North East Europe from which it migrates to warmer places in cold weather. Measuring 170 mm. long the Kingfisher feeds on small fishes and may cause damage in trout hatcheries. It scans the water, either from a perch, or hovering a few metres above the surface and plunges in to catch minnows, bullhead or some other small fish. It usually takes them to an execution post where the fish is banged to death before it is swallowed. The Kingfisher's eggs are laid in a tunnel excavated in a stream bank. No nest material

Stercorarius parasiticus   *Gavia arctica*

is used, but the young eventually lie on a platform of rejected fishbones cemented by their excreta, for Kingfishers are among the few birds which do not practise nest sanitation.

RED BREASTED MERGANSER —
Mergus serrator

The Red Breasted Merganser is a saw-billed duck, characterised by a large crest in both sexes and, in the male, by a reddish-brown patch of colour on the chest, although this is absent in the female. These ducks, which measure about 580 mm. in length and have a wingspan of about 840 mm. nest in sheltered places among rocks or in dense vegetation beside the sea and on the banks of rivers and lakes in the north of mainland Europe, Scotland, Ireland and Iceland. In Winter they migrate, often in small flocks, along the sea coasts as far south as the Mediterranean. Their food is almost entirely fish and this makes them unpoplular with anglers. They are among the few birds not protected by law in Scotland.

ARCTIC SKUA —
Stercorarius parasiticus

This brown gull-like bird, 460 mm. long, nests in Iceland and on the coasts of Northern Europe

and Scotland. It winters south to the north coast of France and may then be seen in estuarine areas. It is a pirate, rarely hunting for itself, but having an uncanny knowledge of any bird which has had some success in fishing. It will chase and harry such a bird until the latter finally disgorges its last meal which will be pounced upon and eaten by the Skua.

BLACK THROATED DIVER —
Gavia arctica

The Black Throated Diver is another northern bird which nests in Scotland, Scandinavia and Northern Russia, usually by lakes in mountains or tundra. In wintertime it migrates as far south as the Mediterranean and may be seen on rivers or inland lakes. At this period, however, the birds have moulted the breeding dress of stripes on the neck and back and appear grey above and white below. In spite of its size, (625 mm. long with a wingspan of 1,100 mm.) it may be difficult to see, for it is very wary and can submerge stealthily and swim under water to a place of greater safety. Its food includes fishes, crustaceans and some plants.

151

Castor fiber

Neomys fodiens

Myocaster coypus

Lutra Lutra

MAMMALS

BEAVER —
Castor fiber

The Beaver is the largest European rodent and one of the rarest, having been hunted to extinction because of its fur over most of the forested areas of the continent where it once lived. The last British Beavers were killed in the thirteenth century and the species now survives in isolated areas on the Rhone, the Elbe, in Scandinavia and Western Russia, although some attempts are being made to reintroduce them. The southern Beavers are somewhat bigger than those from the north, with a total length of about 1,200 mm. They are all extremely shy and wary and mainly nocturnal. They do not normally build lodges like the American Beavers which are members of the same species, although they may do so if sufficiently undisturbed. They usually live in long tunnels which open below the surface of the water, although this may be kept at the right level by the construction of dams. Beavers mate for life and live in family groups which are unusual in that the one litter produced each year takes two years to mature and become independent. Their food includes many sorts of water plants and the bark of small trees, especially willows, alders and birches.

WATER SHREW —
Neomys fodiens

The Water Shrew is to be found throughout Europe except for Ireland, Spain and the Balkan Peninsular. Its total length rarely exceeds 165 mm. and like the other European shrews it is a tireless hunter by day and night of other small animals. These they may subdue by means of a poisonous secretion which enables them to tackle prey up to the size of small mice. In spite of its name and the fact that the keeled tail and fringed feet fit this animal for life in the water, Water Shrews are often found well away from any sort of damp area, although slow-flowing streams seem to be their favourite habitat. When they swim and hunt under water, their velvety coats trap countless air bubbles, which give them a silvery appearance. Like all shrews, this species is short lived, seldom surviving for more than a year and a half. Breeding may occur in the year of birth; the young, numbering up to ten, are born in a nest cavity which may be excavated by the mother. Water Shrews are more social than other shrews and are often very noisy, making squeaking and whistling sounds.

WATER VOLE —
Arvicola terrestris

The Water Vole is often referred to in Britain as the Water Rat, but it is not in fact a true rat, from which it differs in its short face, small eyes and ears and hairy tail. It is a widespread species occurring from Western Europe to central Siberia, although it is absent from Ireland. There is much local variation in size and general appearance in Voles from different areas. Those in Britain are large, measuring up to 360 mm. overall, those from further south may be much smaller and have projecting incisor teeth. The northern and British forms are largely aquatic; the more southerly ones less so and live in tunnels away from the water. The northern Water Vole, which is mainly nocturnal, lives in river banks and is a good swimmer. It slips directly into the water via submerged entrances through its tunnels. Although it may occasionally feed on carrion fish, its food consists chiefly of water plants and it may make stores of these in its extensive tunnel system. Breeding occurs through the summer months and the young, born early in the year, will be breeding also by the end of the season. The enemies of Water Voles include mink, otters and flesh eating waterside birds as well as the more voracious of the fishes which may catch a few. Man, however, is the most important foe, for in Eastern Europe Water Voles are trapped for their fur.

COYPU —
Myocastor coypus

Coypu, which are related to the porcupine, were brought into Europe from their native South America for the sake of their fine fur, which is known as Nutria. They have escaped and established themselves in various places including the Fenlands of England. Here efforts are now being made to keep the numbers to a very low level, for these animals, with a total body and tail length which may exceed 1,000 mm., are destructive to river banks with their large burrows. They are known to have done substantial damage to crops in riverside fields.

OTTER —
Lutra lutra

The Otter is an aquatic member of the Weasel family, adapted by its thick fur and webbed toes to a life in the water. This animal, which may grow to over 800 mm. body length, plus 500 mm. tail length, may be encountered by streams or lakes over almost the whole of Europe. It is, however, nowhere common for its depredation on fishes, even though these may often have been species not wanted by anglers, or sickly specimens, led, nontheless, to its extermination in many areas. In some places it is now officially protected by law and its numbers are increasing, although in Britain this is not yet so. Otters are mainly nocturnal in their hunting and in the long journeys which they may make between one safe area and another. Young may be born at almost any time of the year and seem at first reluctant to enter the water. All young carnivores are playful and adult Otters retain this characteristic.

ARABLE
AND
PASTURE
LANDS

INTRODUCTION

Although forests originally covered much of Europe, these have now largely been destroyed and in their place we find open fields which may be pastures or growing food crops of all sorts. In Britain this destruction was started in prehistoric times and trees were cut and burned to release the land for agriculture and in later times this continued because of the large scale need for fuel. There is no doubt that were man's activities to cease, most of Britain and much of Europe would quickly become reafforested. The fields now look permanent, but they are kept in this state by cultivation, or by the constant grazing by animals, and the only reminder that we have of the forests is in the trees of hedgerows and field margins while the herbaceous plants of the forest edges now survive as weeds. Some woodland animals have managed to adapt themselves to a life in open country with minimal cover. In other cases the felling of forests has offered new living space to creatures of natural grasslands and in mainland Europe these occassionally irrupt into populations of pest or plague proportions. A case in point is the European Hamster (see page 176) — a larger animal than the pet Golden Hamster. In especially dry years in Central Europe enough of the young produced may survive to become a serious pest to farmers. Field Mice and other small rodents in spite of their size may also be pests for they breed so fast that plague numbers can build up very quickly and their depredations on arable crops can be considerable. Here, predatory birds and mammals are man's chief friends, for the little rodents are their main food and when rodent numbers are high the breeding success of owls and small hawks such as the Kestrel is assured and they with foxes, stoats and weasels help to keep the plague in check. Rabbits and hares may also become pests, although since both are good eating they are used as food animals by humans and this predation may keep their numbers low. In some areas the fur as well as the flesh is valued. The numbers of rabbits have declined drastically in many areas since the introduction of myxomatosis in the early 1950s. This disease, which is endemic to South America, where it affects rabbits, has barely more effect on them than the common cold on an Englishman, and recovery is usually rapid and total. The virus when introduced to Australia proved to be devastating with a high mortality resulting from infection. The same occurred in Europe and although in Britain at least some rabbits now seem to have a partial immunity, the disease has become endemic — rabbit populations are periodically infected and a number die as a result.

Cultivation opens the door to many kinds of plant eating insects, which may become abundant in farmland, often damaging crops or pastures during their larval phases. Occasionally great plagues may build up but normally they are kept in check by predators including many bird species and some insects such as the Golden Ground Beetle (see page 163). The great

French naturalist Fabre, describing the onslaught made by these creatures on some caterpillars, said that their ruthlessness and efficiency was reminiscent of nothing less than the great stockyards of Chicago.

In highland Europe small fields have often been cleared on the slopes of mountains. These, in terms of modern large-scale agriculture, are not efficient units, but they are often vital to peasant communities. Besides this they frequently offer great natural history interest in the plants and animals which occur there, which may be different from the flora and fauna of the surrounding areas. In places where this sort of agriculture is now being abandoned, these areas are sometimes declared as nature reserves and act as reservoirs of many organisms which are very rare elsewhere.

MOLLUSCS

Abida frumentum

This little snail, which rarely reaches a size of more than 6—10 mm. high, is an inhabitant of Europe south of the Alps and in Spain. It is to be found on walls and rock outcrops, feeding on the small plants which grow in such an environment.

GARDEN SNAIL —
Cepea hortensis and
GROVE SNAIL —
Cepea nemoralis

These two species of closely related snails are to be found widely over Europe, although the Garden Snail occurs in slightly more northerly and westerly areas and shows some preference for taller vegetation. They are about the same size with a shell height of up to 16 mm. and a width of 24 mm. and are both exceedingly variable in coloration. This ranges from pale yellow to pink or brown which may be clear or banded with a variable series of brown stripes. The genetics of this is well understood and is one of the classic cases of a polymorphism held in balance by the selective predation of thrushes. The species can be generally distinguished by the possession of a dark rim to the tip of the lip of the Grove Snail which is absent in the Garden Snail.

INSECTS

GREAT GREEN BUSH CRICKET —
Tettigonia viridissima

This insect is found in the southern counties of England and Wales and widely over Europe and temperate Asia. It is the largest member of the grasshopper family native to Britain, for the male measures 40 mm. long and the female, with her curved ovipositor is larger still. In spite of its size, the Great Green Bush Cricket can be difficult to see for its colour merges perfectly with that of the vegetation on which it rests. It can, however, often be pinpointed

by the shrill song of the male, which starts in the early afternoon and often carries on well into the night throughout the summer months when the birds are active. The eggs are laid in the soil, and the nymphs when they hatch feed, as do the adults, on a wide range of plants and also on other insects. If disturbed these insects usually walk rather than hop, and in spite of the size of their wings, the adults rarely fly far.

FIELD CRICKET —
Gryllus campestris

The shrill chirrup of the male Field Cricket is a familiar sound in much of Southern and South East Europe although it is becoming increasingly rare in Britain, where the species now survives in only a few southern counties. The sound is not produced vocally, but by rubbing wings together on which there are specially roughened veins and producing a high pitched whistling noise. Although the singer is about

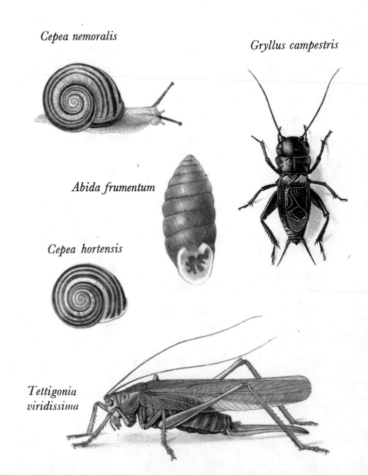

Cepea nemoralis

Gryllus campestris

Abida frumentum

Cepea hortensis

Tettigonia viridissima

20 mm. in length it is far from easy to find, for it sits at the mouth of a burrow into which it retreats at the least disturbance. The females, which do not sing, occupy burrows during the winter months but leave these to find a mate whose home they then share.

After mating the females lay up to about 300 eggs in the soil. The nymphs which hatch from these do not complete their growth until the next year, overwintering in their burrows. Field Crickets feed on plants and other animals but in captivity they can survive on a relatively restricted diet of leaves, although they must have some water to drink as well. In some Southern European countries it is still possible to buy tiny cages in which these singing insects may be kept, although nowadays they are usually made of plastic rather than the elegantly carved wood decorated with silver bells which used to be common.

SQUASH BUG —
Coreus marginatus

This insect gets its odd name from a close relative which infests squashes in America, although *Coreus marginatus* feeds on members of the dock family. It is found in Britain from the Midlands southwards, and widely in Europe. The nymphs hatch in the midsummer months and feed on their host plants, moving in Autumn to ripening seeds as food. They then overwinter before becoming sexually mature but complete their development early the next Summer.

Polistes gallicus

Sheltered spots in fields in Southern Europe are the usual home for *Polistes gallicus*, a wasp which does not occur in Britain, but may be found in the warmer parts of the Continent, in North Africa and across Asia to West Pakistan. The colonies are small, and the nest made of paper, as in all wasps, is distinctive because it is not completely enclosed in a covering envelope. The colony consists of a queen which is the mother of all the workers which are sexless females, drones which are males and unmated females.

As with all wasps, the grubs, reared by the workers, are fed on flesh, mainly that of other insects.

BEDEGUAR GALL WASP —
Diplolepis rosae

During the summer months, wild roses may often be seen with a number of fluffy red tufts growing from the twigs. These galls, sometimes called Robins' Pincushions, are the result of a tiny wasp, only 4—5 mm. long, having laid her egg in the rose plant. At the point of injury, the plant reacts with a growth of abnormal tissue, which provides protection and food for the grubs of the wasp, each gall containing several cells with developing larvae. They are full grown by late Summer and rest through the wintertime, pupating in early Spring. Males are known from this species, but the vast majority of individuals are females, which can lay fertile eggs without mating.

LEAF CUTTER BEE —
Megachile centuncularis

In gardens and field edges plants, particularly roses, may sometimes be seen to have neat round holes cut out of the leaves or petals. This is likely to be the work of a Leaf Cutter Bee, which is widespread through Britain and Europe. The female bee which is about the size of a Worker Honey Bee but dark coloured, tunnels into a rotten stump or fence post often to a depth of 30 cm. or more. The pieces of leaf or flower petal which she cuts are carefully arranged in this to make a series of thimble-shaped cells, each provisioned with pollen and each containing an egg. The bottom cell is complete first, yet the bee which develops in it will hatch last so that its emergence will not disturb its sisters.

BUFF-TAILED BUMBLE BEE —
Bombus terrestris

The illustration of this large Bumble Bee shows the continental form in which the tail is pure white; in Britain this species has, as its name suggests, a yellowish coloured tail. It is a common species through most of Europe other than Scandinavia and is particularly abundant in the south and in North Africa. The life history is similar to that of other Bumble Bees. The fully sexed females mate towards the end of the Summer and then go into hibernation. Next Spring they seek out a suitable underground hole, often in an old mouse nest, which may be approached by a tunnel over 1,5 metres long. Here, in the remains of the nest lining left by the mouse, she makes a small honey store and lays eggs on a mound of pollen which she has collected. The grubs which soon hatch from the eggs are tended by their mother and when they finally attain adult form they take over most of the work of the nest, leaving the queen the task of further egg laying. A large colony may contain 300—400 sexless worker bees, but these all die at the end of the Summer. Bumble Bees have very long tongues and are important pollinators especially of clover.

TIGER BEETLE —
Cicendela campestris

This Tiger Beetle is found generally throughout Europe on heathland and into Western Asia. Although only about 16 mm. long it is a voracious carnivore, feeding on any small animal which it can catch on the ground, but flying very readily for short distances if frightened or disturbed. The eyes of the Tiger Beetle are large, and it spots the ants which are its main food from a considerable distance — up to 15 centimetres away — and rushes in to grasp them with its huge mandibles. However, other creatures such as spiders, caterpillars and even slugs will be caught and sucked dry if the occasion offers. After mating the female Tiger Beetle lays her eggs singly in pits up to about 10 millimetres deep which she digs with her mandibles. She covers each egg carefully with soil so that it is protected against the weather and other creatures which would like to eat it. When the larva hatches it digs a small vertical tunnel in which it sits, wedged near the mouth by a hump on its back which supports it. Any small insect which comes near is seized by the insatiable grub, which drops quickly to the bottom of the tube if danger threatens. The larval life lasts for two to three years, the creature overwintering deep in the soil. Pupation is brief,

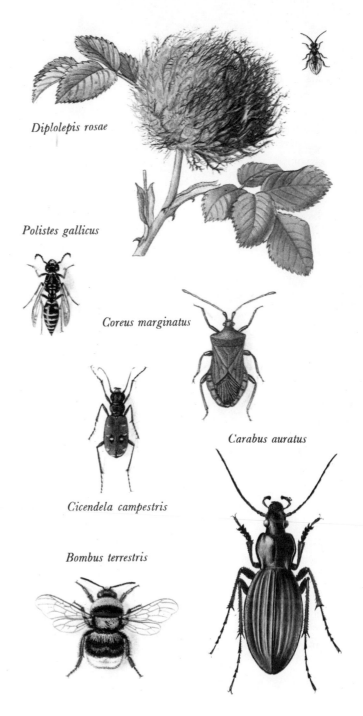

Diplolepis rosae

Polistes gallicus

Coreus marginatus

Carabus auratus

Cicendela campestris

Bombus terrestris

but after emerging in late summer, the adult Tiger Beetle does not breed until the following year.

GOLDEN GROUND BEETLE —
Carabus auratus

This beautiful beetle is widespread in Western Europe,

163

but is not native to Britain, where, if it occurs at all, it is probably imported accidentally with shipments of vegetables. Like the Tiger Beetle it is a hunter, but is largely nocturnal. It is one of the most useful of insects to farmers for one of its main foods is cock-chafers (see page 222) and their eggs, although caterpillars, worms and any other small animal will be attacked by it. Stories of Golden Ground Beetles hunting communally do not seem to be true. Rather, they are solitary animals, but in areas where high populations are present, several may shelter under the same stone or log and may all descend on the same unfortunate prey. Each, however, is working for itself alone and will tear fragments of food from the jaws of the other beetles rather than helping them. Early in the Summer the females lay a rather small number of eggs in the ground. These hatch in slightly over a week into predatory grubs, which are full grown towards the end of the Summer. Pupation is brief and the adult insects emerge in late Summer, but remain underground and do not start their life of butchery until the next Spring.

Zabrus tenebrioides

This beetle, while related to the former species, is not a hunter but a grain feeder and as such is a major pest in some parts of Central, Southern and Eastern Europe, where it is common. In Britain, it is found only in a few areas in the southern counties and is not regarded as particularly harmful. The adult beetle, which is about 14—16 mm. long lies hidden in the ground during the day, but at night feeds on the growing crop. The larvae are grass feeders and are also considered to be agricultural pests.

BOMBARDIER BEETLE —
Brachinus crepitans

Bombardier Beetles are to be found in southern Britain and in Central and Southern Europe, where they may occasionally occur in considerable numbers. These little insects, which are only 6—9 mm. long, have a remarkable form of defence from which they get their name. Just as a predator imagines he has a tasty tit-bit, for an entomologist an interesting specimen, the beetle produces a loud bang, which will startle even a human being and may terrify a smaller attacker. The mechanism of this defence is that the beetle carries in its abdomen two fluids, which are secreted under pressure at moments of stress. As they meet and fuse the small chemical reaction causes the report. The vapour, although invisible, can make a brown stain on a human hand and is doubtless unpleasant in odour or taste to lesser enemies.

CLICK BEETLE —
Agrypnus murinus

Click Beetles are so named from their ability to leap to safety when on their backs with an audible click. The one illustrated is fairly common in southern and central Britain, less so in the north and in Ireland. In Europe it is generally distributed in many sorts of habitat although it becomes rarer in the north. Click Beetles are herbivores feeding on many kinds of vegetation and are often pests. *Agrypnus* adults feed on the foliage of trees and also in cultivated crops. The larvae are even more of a pest, for they live in the soil and damage the roots of young trees.

WIREWORM —
Agriotes lineatus

The Wireworm gets its name from the appearance of the larva, which is an elongated, yellowish coloured grub with a hard cuticle. This lives in the soil for the three years of its larval life, feeding on the roots of a wide range of plants including field crops of all sorts, vines and trees. Although attacked by many predators, both vertebrate and invertebrate, the Wireworm occasionally builds up very large populations which cause great damage to agriculture. The adult is a Click Beetle, 8—9 mm. long, narrow bodied and capable of leaping in the typical manner of its family. This species is widespread in Britain and Europe as far north as central Scandinavia.

Agrilus biguttatus

This 12 mm. long beetle may be found in a few

localities in southern Britain, although it is common in Southern and Central Europe. Its preferred habitat is wooded slopes or fields near to forests. The larvae develop in the stumps of dead oak trees; the adults may be seen sunning themselves in early Summer.

SEVEN-SPOT LADYBIRD —
Coccinella septempunctata

Ladybirds, in spite of their name, are small beetles, well known because of their bright colours and their non-retiring habits. They can afford to behave in this manner because they taste very unpleasant; any predator which tries to eat one will almost certainly reject it, and reminded by the vivid colours, will not attack another. Ladybirds are among the best loved of beetles and in many European languages their name signifies that they are of sacred origin. The reason for this is almost certainly that Ladybirds

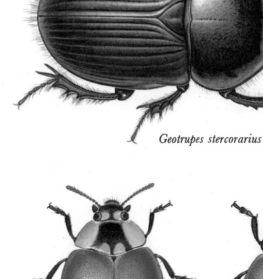

Geotrupes stercorarius

Adalia bipunctata

Coccinella septempunctata

Agrypnus murinus   *Leptinotarsa decemlineata*   *Agriotes lineatus*

Agrilus biguttatus

Brachinus crepitans

Zarbrus tenebrioides

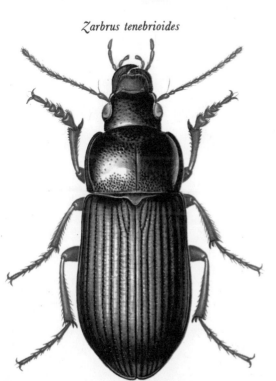
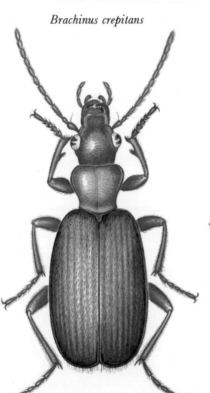

are a most useful insect to man, since their diet is almost exclusively of aphids. The Seven-spot Ladybird which is common and widespread in Britain and occurs through Europe and Asia as far as India, destroys these pests throughout its larval and adult life.

TWO-SPOT LADYBIRD —
Adalia bipunctata

This species of Ladybird is only 4—5 mm. long, but is an avid destroyer of aphids. The larval life lasts about a month, during which time the grub destroys between 200 and 300 and after pupation the adults continue to feed equally voraciously. A female may lay about 200 eggs in her life-time, so it can be seen that these are valuable controls of aphid pests. Two-Spot Ladybirds are common in Britain and Europe and occur also in North America. Many overwinter and sometimes hibernate in large numbers in houses or other sheltered places.

DOR BEETLE —
Geotrupes stercorarius

This beetle which is widespread in Britain and Europe and North Africa occurs also in Newfoundland. It is very closely related to the Dumble Dor (see page 24) which it replaces in agricultural land. The habits of the two species are virtually identical, except that *Geotrupes stercorarius* normally uses cattle or sheep dung to provision its brood chambers.

COLORADO BEETLE —
Leptinotarsa decemlineata

The Colorado Beetle which is about 10 mm. long, originated in Mexico, but spread during the last century into the United States and from there to Europe. Here, with no natural enemies to resist it found a haven and has subsequently become a serious pest of potato crops. Although it occasionally reaches Britain, it apparently cannot survive the wetness of the Winters and has never established itself there. As both larva and adult this beetle is a pest to potato crops, which it can decimate by destroying the green haulms of the growing plants. It develops very quickly during the summer months and several generations may add to the total destruction wrought.

Syntomis phegea

This moth which does not occur in Britain, may occasionally be seen in Western Europe, but is really a native of the southern part of the Continent, where, during June and July, it may be found feeding on the flowers of thyme and lavender. Dandelions are the favourite food plants of the caterpillars, which are recognisable by the feathery black hairs along the back. They overwinter as larvae and complete their development early the following year.

Lemonia dumi

Lemonia dumi, which has a wingspan of about 50 mm., is a moth of the autumn months, and may sometimes be seen as late in the year as November. It is found over the whole of Europe other than Britain. The larvae which are dark brown with satiny black patches along their sides, feed on dandelions, hawk weed and related plants including lettuces, but the species is never common enough to become a pest.

TURNIP MOTH —
Agrotis segetum

This species is to be found in early Summer throughout almost all of Britain and over much of Europe and temperate Asia. It has a wingspan of about 40 mm., but its colour range is very great although the hindwings are always paler than the forewings. These moths differ so much from one individual to another that in the past several species have been described based on such variation. The caterpillar is greyish brown with dark spots and light stripes on its back and sides. It feeds on a range of field crops, especially turnips and swedes, usually attacking the stems just above the ground. It overwinters in the shelter of a cavity which it gnaws in a root and pupates in the Spring, emerging in early Summer.

166

Herse convolvuli

Agrotis segetum

Lemonia dumi

Syntomis phegea

Acherontia atropos

CONVOLVULUS HAWK MOTH —
Herse convolvuli

The Convolvulus Hawk Moth is to be found throughout much of the Old World, Australia and Polynesia. In the warmer parts of its range it has no special breeding season and several generations are produced

167

Hesperia comma

Gonepteryx rhamni

Eristalis arbustorum

Vanessa carduii

Mayetiola destructor

Pieris rapae

each year. These may migrate away from areas where the species is abundant, often into places where breeding is unlikely to be successful. A few reach Britain in most years, and occasionally there are vast irruptions of the species, but successful breeding is rare so far north although eggs are often laid, usually on lesser bindweed in potato fields. During the daytime the moth rests, but is active at night, its fast wingbeat making a distinctive whirring noise which may be the first clue to its presence, for in spite of its size (115 mm. wingspan) it is difficult to see. It feeds on flowers such as *Nico-teana* which have nectar at the base of a long tube. This is reached by the moth's tongue, which may measure nearly 130 mm. in length.

DEATH'S HEAD HAWK MOTH —
Acherontia atropos

The bulkiest of the European Hawk Moths, the Death's Head has a wingspan of 115 mm. It is really a creature of warm climates, occurring in Africa and parts of Asia and migrating to Europe. A few reach Britain in most years and occasionally very large numbers are recorded. They usually lay their eggs on potato plants although these cannot have been the original food as potatoes are American in origin. Eggs

may be laid on a variety of other plants, including jasmine and nightshades. The caterpillar grows to a length of 130 mm. before pupation which is in the ground. Features which distinguish the Death's Head, other than its large size, are the 'skull' markings on the thorax, the large eyes, which are said to glow at night and the fact that it can make a loud mouse-like squeak. It has a short tongue and rarely if ever feeds on flowers, but is attracted to bee hives and may raid these for the honey.

SILVER-SPOTTED SKIPPER —
Hesperia comma

The Silver-spotted Skipper may be seen in July and August over the chalk hills of southern England which is the only place in Britain where it is to be found. It is widely but patchily distributed through Europe and Asia and western North America, though not in the hottest regions and is usually associated with a calcareous soil, even when this is quite high, for they are recorded as occurring at over 2,000 metres in mountainous country. The female, with a wingspan of 30 mm., is larger than the male, which is illustrated here. The caterpillars feed on grasses which they attack near to the base of the stem. Here they protect themselves by spinning the fine leaves together with silk. Subsequently they pupate near to the ground.

PAINTED LADY —
Vanessa carduii

The Painted Lady is found in every continent except South America, although those seen in Europe are mainly migrants from Africa which may occasionally reach as far north as Iceland. The food plants are thistles and nettles, but in Britain, even though it may breed successfully the species is unable to over-winter and butterflies seen in the Spring are migrants, capable in spite of a wingspan of only 50 mm. of making the long journey probably from Africa.

SMALL WHITE BUTTERFLY —
Pieris rapae

Small White Butterflies may be seen throughout

Europe and across Asia to Japan and have been introduced into North America. In Britain the numbers are increased by migration from the Continent, although it is able to breed successfully, the larvae feeding mainly on plants of the cabbage family. Two generations are normally produced in a year; the second, which is illustrated here, have more black on the wings than the spring brood.

BRIMSTONE —
Gonepteryx rhamni

This butterfly which is found from North Africa to Northern Europe and through much of Asia to Siberia is, in Britain, one of the creatures that herald the Spring, for the adults hibernate but often fly in the warmth of early sunshine. The food plant is buckthorn, and the caterpillars which hatch from the eggs laid by overwintering insects feed, pupate and are themselves adult by late July. They hibernate in the Autumn and reappear and breed next Spring, thus having a longer adult life than the majority of British butterflies.

DRONE FLY —
Eristalis arbustorum

This hover fly, which has a wingspan of about 15 mm., may be seen through much of Britain and Europe in the late summer months, feeding on such flowers as ragwort. In spite of the resemblance to a small Bumble Bee, which extends even to the noise of the wingbeats, it is quite harmless and unable to sting.

The eggs are laid in water, often in ponds or ditches which are very highly polluted, but the larvae which are called Rat-tailed Maggots survive by means of telescopic breathing tubes which they push to the surface of the water when they need air.

HESSIAN FLY —
Mayetiola destructor

This wheat midge, only about 3 mm. long, gets its name because it is supposed to have been taken from Europe, where it is common, to America by Hessian troops in the eighteenth century. Over much of the warmer parts of its range it is very destructive to wheat crops for the larvae bore into and cause galls on the stems, which frequently die as a result. In Britain, however, it seems to do little damage although it is widespread.

BIRDS

ROOK —
Corvus frugilegus

Rooks are resident in Britain and through most of Europe where the ground is unfrozen in Winter. They spread north of this line in the summertime

Corvus frugilegus

Pica pica

but retreat to central and southern parts of the Continent as the weather hardens. Although large (460 mm. long) they may be difficult to distinguish from their close relative the Carrion Crow (see page 42) although they may usually be recognised by their glossy black plumage and the rather loose 'trouser' feathers. At close quarters the white skin at the base of the bill and relatively slender beak will be further distinguishing marks. Rooks are social birds, nesting communally in high trees in early Spring and feeding in flocks later in the year, often in company with jackdaws and starlings during the winter months. They eat a wide range of foods, both plant and animal. At times they are a nuisance to the farmer; at others, when feeding on Wireworms and similar pests they are beneficial. Rooks' nests are solid structures made of sticks and lined with grass and moss and are refurbished and used for many years. The four to five eggs are incubated by the female for nineteen days during which time she is fed by the male. Later they both tend the young, which at first do not have the bare facial skin and so may be difficult to distinguish from crows.

MAGPIE —
Pica pica

The Magpie has a total length of 460 mm., equal to that of the Rook. It is, however, a much smaller bird, for half of this is taken up by its long tail. It is to be found throughout the whole of Europe where ever fields or open spaces are interspersed with scrubland or dense trees and bushes. Although common it is very wary for in many places man's enmity towards it is constant. It is, for example, one of the few unprotected birds in Britain. The reason for this is its liking for the eggs and young of other birds whose nests it pillages during the springtime, often damaging populations of song or game birds very seriously. In spite of this persecution the Magpie holds its own and its scolding chatter is a familiar sound of the countryside. Magpies' nests are solid, domed structures, usually built in a dense thicket. They are not reused in subsequent years, although they may be taken over by owls or birds of prey. The eggs are incubated by the female,

but the young are cared for by both parents with whom they remain for some time after fledging. Small parties of Magpies may be seen therefore in late Summer although large flocks are rare.

SKYLARK —
Alauda arvensis

Skylarks are to be found over the greater part of Europe during the summer months, although in Winter they migrate south and west from the colder parts of the Continent, some travelling as far as Africa and South West Asia. It is essentially a bird of open country, where from early Spring onwards the beautiful, varied song of the male may be heard. This is given in a vertical flight, over the territory, the bird often ascending to a great height, rising and falling in the air as it sings. The nest, which is beautifully concealed, is in a small depression on the ground, and here the female incubates three to five eggs for about thirteen days. The young leave the nest before they are fully fledged and are cared for by their parents for some while until they are quite able to fly and find their own food. Skylarks feed on insects, worms, grubs, spiders and small seeds which they find in open fields. They are regarded as gastronomic delicacies in some parts of Europe and are eaten in large numbers. This habit is now decreasing and in many countries, as in Britain, both the adult birds and the nest are protected.

TREE SPARROW —
Passer montanus

The Tree Sparrow is found over all of Europe except for the coldest parts of the north, parts of the Balkans and much of Ireland. Both sexes are alike, measuring about 140 mm. long and in general appearance are very much like the House Sparrow, but the head is chocolate brown and there is a black spot on the white patch behind the eye. It is more of a country bird than the House Sparrow and is commonest in cultivated areas where there are plenty of trees. These are necessary to it for it nests in holes in tree trunks, often producing three broods in a season, feeding the young chiefly on insects. But open country is

170

Passer montanus

Alauda arvensis

Emberiza citrinella

Acanthis cannabina

vital also for the adult feeds on grain and seeds and some insects which it gets from the fields.

YELLOWHAMMER —
Emberiza citrinella

The cock Yellowhammer is one of Europe's most brightly coloured birds and is often mistaken for an escaped cage bird, although if he is singing his territorial song in the spring and summer months there can be no doubt as to his identity. This song, usually rendered as 'a little bit of bread and no cheeese' is repeated endlessly from a low bush or small tree in a shrubby area where the nest is made on the ground. The female bird, which like her mate measures about 165 mm. long, is far less brightly coloured in dull shades of brown. Yellowhammers are to be found through most of Europe, although they do not nest in the extreme south and east, and retreat in Winter from the most northerly areas. After raising two or sometimes three broods during the summertime, Yellowhammers flock up, often with other species, which roam the open countryside looking for the seeds and occasional small animals on which they feed.

LINNET —
Acanthis cannabina

Linnets are found through much of Europe during the summer months, although they migrate southwards from the coldest areas during the wintertime. They are birds of open country especially in Winter where there are bushes and hedges in which they can nest.

They also like heats, parks and gardens. Both male and female are about 130 mm. long, but he is much more brightly coloured than his mate who lacks any pink on the breast or the top of the head. The courtship and territorial song of the male Linnet is a very pleasant musical twittering which is very variable from one bird to another and is often given from the perch. In the past in Britain and still in some parts of the world where Linnets occur they are favourite cage birds, for their song is highly adaptable and they can be taught tunes very different from their natural repertoire.

The nest is built and the eggs incubated by the female, although the male helps to rear the nestlings. These are fed at first entirely on small insects, although as they grow their diet changes to include a high proportion of seeds, mostly of various species of agricultural weeds. In Autumn, Linnets and their young form mixed flocks with other species of finch, scouring the fields and banks for food. Over much of its range the Linnet seems to be scarcer than formerly, possibly because its food plants have diminished and perhaps also because of the effect of insecticides on the young.

RED-BACKED SHRIKE —
Lanius collurio

In Britain the Red-backed Shrike occurs as a nesting species only in south east England although it is found in north Spain across Europe to Japan. It is, however, in all cases a migrant, departing for the tropics at the end of the Summer. It is a bird of open areas where there are dense bushes in which the nest can be built. It can be recognised as a Shrike by its rather thick-set appearance (it is about 170 mm. long) with a heavy beak, hooked at the tip, and a long tail. When scanning the surroundings for food it may sit with a characteristically upright stance on an exposed branch or on a wire fence, although it may hover by hedges or near to cover where food is likely to be found. All sorts of animal food is taken from large insects to young birds, small mammals and reptiles. Food not due to be eaten at once is often impaled on a thorn or barbed wire in a larder — a habit common to all Shrikes which is the reason for their being called Butcher Birds. The male helps to build the nest, but incubation is mainly by the female although both parents feed the nestlings and continue to care for the young after they are fledged. The normal cry of the Red-backed Shrike is a harsh squawk, but the song of the male, usually sung from a high perch, is musical and varied.

GREAT GREY SHRIKE —
Lanius excubitor

The Great Grey Shrike is resident over much of Central and Eastern Europe and Spain and nests as far north as Scandinavia, though it migrates from the colder parts of its range in wintertime to the south and west of the continent. In Britain it is seen mainly as a winter visitor and does not stay to nest. It is the largest of the Shrikes, measuring 240 mm. long, but is generally similar in habits to the other members of its family, though it inhabits country with denser scrub than the other species. The flight is normally fairly close to the ground and undulating, often ending with an upward swoop to the perch, though when hunting it frequently hovers as an alternative to sitting in a tall tree or on telegraph wires to examine the ground for food. The sounds made by the Great Grey Shrike include Magpie-like chattering and the song intersperses these harsh calls with more subdued, musical notes.

WHITETHROAT —
Sylvia communis

One of the commonest summer migrants to lowland farmlands, the Whitethroat is found in suitable habitats over almost the whole of Britain and Europe as far north as southern Scandinavia. Its scolding song may be heard from the thickets or nettle patches where it makes its nest, or occasionally in a brief song-flight. Both parent birds incubate the eggs and feed the young on insects, grubs and spiders even after they have left the nest. When breeding is completed the Whitethroats, although only 140 mm. long, start the journey back to Africa where they spend the winter months.

KESTREL —
Falco tinnunculus

The Kestrel is the most widespread and the most

Lanius collurio

Sylvia communis

Lanius excubitor

Otis tarda

Falco tinnunculus

common of the European birds of prey. Being small (only 340 mm. long at most) it is regarded with less suspicion than its larger relatives and is often unmolested. It lives in almost any open country where nest sites in the form of hollow trees or cliff faces or even nest boxes are to be found. It is one of the few carnivorous birds which has learned not to fear man, and may nest in the artificial cliffs of big buildings in towns. There are several pairs which breed regularly in London and other British cities. Its food includes any small rodent or large insect which it can catch on the ground. It has the characteristic habit of hovering about 10 metres above ground, its tail fanned out and flickering to maintain its balance while it scans the area for food. This flight is the reason for one of its old country names 'the Windhover' for it appears to be standing on the air. Food is usually swallowed whole and pellets of indigestable material ejected later. The specimen illustrated is a male. The female, which is slightly larger than her mate, lacks the grey head colour and is speckled brown there as on the back and breast.

GREAT BUSTARD —
Otis tarda

At one time Great Bustards roamed the open chalk uplands of south central England, but they have long since become extinct there as in most other areas in Europe. Populations remain only in southern Spain, north central Germany and South East Europe and across on to the steppes of South West Asia. Their large size and reluctance to fly made them easy targets for hunters in spite of their extreme wariness. Great Bustards normally live in flocks, in which females which are very much smaller than their mates predominate. The size difference is an immediate method of distinguishing the sexes, for the males are about 1,020 mm. long and have a wingspan of about 2,205 mm., while the females are 760 mm. long and have a wingspan of 1,780 mm. In the breeding season the males play no part in incubation or rearing the young, which are fed by their mothers at first on insects and other small animals but later become mainly grazing creatures. An attempt is being made to reintroduce these birds to some of the ancient English haunts.

PARTRIDGE —
Perdix perdix

The Partridge is resident through most of Britain and much of mainland Europe, other than the far north and Spain. It is a stumpy looking bird, about 300 mm. long, which likes to have thick hedges or scrub where it can nest in safety, bounding the fields where it feeds. Recent agricultural trends in Britain which involve the removal of many hedges has reduced

Perdix perdix

Coturnix coturnix

the suitable living places for the Partridge. The nest is made on the ground, and the chicks suffer if, as seems to have happened recently, there is a cold or wet spell in late Spring. These two factors have led to a decline in Partridge numbers, in spite of the fact that the females lay large numbers of eggs. The young are fed at first on insects, including aphids and the larvae of sawflies; later their diet is almost exclusively of grain. Young Partridges remain with their parents into the winter months and leap almost vertically into flight, their short wings whirring, at the approach of a human or other predator.

QUAIL —
Coturnix coturnix

Quail are the smallest European game birds, measuring only 180 mm. long, and looking rather like miniature Partridges, to which they are related. At one time they were abundant in open country throughout much of Europe, but unlike other game birds which fly fast, but rarely far, Quail are migrants, travelling to Southern Europe, Africa and parts of South Western Asia in the Autumn. They had always been trapped for food while on these long flights, but in the early years of this century the exploitation was increased to such an extent that the Quail as a European species declined nearly to vanishing point. Nowadays a few nest in southern England, feeding mainly on insects found in fields and among crops. It is generally a solitary bird, except when on migration when it may be recognised by its fast wingbeat and low flight.

MAMMALS

BROWN HARE —
Lepus capensis

Brown Hares are found through most of Europe other than the east and extreme north, where they are, however, extending their range. They also occur on the steppes of Asia as far as China and in suitable country throughout Africa. The size is somewhat variable and in Europe it may grow to a total length

of 685 mm. although the Mediterranean race is considerably smaller. Hares are solitary creatures of open country; they do not burrow like rabbits but take shelter in slight depressions in the ground. These are called forms and the Hare which is mainly nocturnal lies out during the day, protected by its stillness and the brown colour of its fur which merges with its surroundings. The breeding season is a prolonged one, starting in early Spring. It is at this time that the 'mad March Hares' are in evidence and they may sometimes be seen chasing and jumping and

Cricetus cricetus

Citellus citellus

Lepus capensis

Microtus arvalis

standing on their hind legs boxing with each other in what are probably complex territorial boundary displays. The young which are called leverets are born at an advanced state with fur and open eyes. Their mother puts them in separate safe places in a field, where they stay very still, almost whatever danger appears, until she comes to feed them. They can run within a day and are weaned in little more than a week and are fully independent soon afterwards. Up to four litters may be produced in a season, but the young do not breed until the following year. Hares feed on grass, field crops and in hard weather on the bark of trees or on garden plants. If numerous they can do a great deal of damage, but their numbers are kept in check by many natural enemies and by human predation.

EUROPEAN GROUND SQUIRREL —
Citellus citellus

The European Ground Squirrel is an inhabitant of dry steppe country in the south east of the Continent and in plains country from Austria to Poland. In these areas it to a large extent replaces the rabbit as a medium sized plant eating social burrower (body length about 200 mm. plus tail about 65 mm.). Each animal has its own tunnel in the colony where it rests at night, but from which it will emerge to feed during the daytime. Ground Squirrels have large cheek pouches and gather food, especially grain, which they store underground. Those which live in cultivated regions often do a great deal of damage to crops and are regarded as agricultural pests. They hibernate through the winter months and breed soon after they wake in the springtime. Only one litter is produced in a season and the young mature rather slowly, not breeding until they are a year old. Ground Squirrels make a wide range of warning sounds.

COMMON HAMSTER —
Cricetus cricetus

This animal is not the wild form of the popular pet hamster, but is a larger creature, measuring about 300 mm. body length plus a tail length of about 40 mm. It is found in the damper, open steppe and plains country from Belgium to Siberia and from the north German plain to the Caucasus. It leads a solitary life in a complex of tunnels which contain nest and food storage chambers. It is almost entirely nocturnal in its activity, and unlike the Ground Squirrels usually produces two litters of six to twelve young a year. These are quickly independent but do not, even so, breed until the following year. Hamsters feed mainly on seeds, but also on other vegetation and on insects. In the late Summer it makes considerable stores, often of cultivated grain, in its burrows and its activity in this respect has led it to be considered an agricultural pest. During the Winter it hibernates but not as deeply as many animals for it wakes quite frequently and feeds on its stores. At all times Hamsters seem to be aggressive and quarrelsome. Even during the mating season males and females keep to their own burrows.

COMMON VOLE —
Microtus arvalis

The Common Vole is to be found over much of Europe other than Scandinavia and the north east, Spain and Italy and Britain, other than forms which occur in Orkney and Guernsey and were probably introduced by humans in prehistoric times. Its size is somewhat variable but may go up to a total length of 135 mm. (including 75 mm. tail length). It is a creature of grasslands, able to make use of the grazed fields because it makes extensive burrows just below the surface. Here it rests through most of the day, becoming active at dusk, when it goes out to feed, mainly on grasses although during the wintertime especially it may eat root crops and take the bark from trees. Common Voles themselves are a major food of many predators including foxes, stoats and weasels. Breeding occurs throughout the summer months, each female producing a series of litters of from three to six young, which may be reproducing themselves at the age of six weeks. Periodic fluctuations of population occur, sometimes building up to plague proportions of several hundred Voles to the hectare, at which time they become an agricultural pest, although a great build up is always followed by a dramatic crash of numbers.

MEADOWLANDS

INTRODUCTION

Meadows may be defined as areas of grassland, sometimes artificially maintained, beside rivers or lakes and on high Alpine slopes. In some cases they are natural grassland; in others they have been created from some other type of plant cover. They differ from the grassy plains of the open steppes in that they are moister and thus produce a lusher type of vegetation. They need relatively little maintenance apart from mowing for hay, or silage or grazing, with cattle in the lowlands and cattle or sheep in upland areas. The droppings of these animals manures the ground, but in hayfields the soil needs to be refreshed with fertilisers, these days usually of a type which replaces lost nitrogen or phosphates. In some meadows close to rivers flooding takes place each Spring as snow melts in the mountains and swells the volume of water which spills over on to the flat country of the lower course. The silt carried by the river is often rich in minerals, and this, which is spread by the flood water, acts as a natural fertiliser.

Apart from grasses natural meadows contain many species of flowering plant, mostly perennials which can stand mowing. Marsh Marigold in the springtime and Ragged Robin later in the year are two well known examples, and although in intensively farmed areas herbicides may be used to eradicate them, in some places they are still abundant. The flower filled meadow attracts many insects, of which the most noticeable are the butterflies. In Britain these have been greatly reduced in recent years by the use of herbicides which kill their food plants (often agricultural weeds) and by insecticides, usually intended for pests of crops but which are unable to distinguish between one insect and the next. In parts of Europe where modern chemical farming is not yet commonplace, the profusion of butterflies in summertime comes as a revelation to visitors from more intensively farmed areas. Several species of Blues, Coppers, Skippers, Fritillaries, and in Southern Europe, Clouded Yellows may be abundant, and moths such as the Burnet Moth may add to the colourful scene. Hoverflies and beetles, some of which are brightly coloured may also be present in considerable numbers, often feeding on the flowers of umbellifers or composites. Bugs, including the Froghopper, which is disguised in its early life in the froth of Cuckoo spit, are common as are Grasshoppers of many kinds, whose ceaseless chirping replaces bird song during the summer months. The short horned species sing mainly by day and the long horned Grasshoppers or bush crickets chiefly at night.

Although many birds may be seen occasionally in meadowland, there are few which are adapted specially to this environment. Hoopoes may be seen probing the ground with their long beaks, and Storks stride by the riverside in their search for insects, amphibians or small rodents for food. Flesh eating birds, such as the Hen Harrier and the Short Eared Owl, both of which have a buoyant flight, may be seen low over the grass looking for suitable prey and they may nest and raise their broods there. Large mammals may pass through the meadows to drink or like deer, come to them to graze at night. Small mammals include the Field Vole, often in very large numbers. This animal, more than any other of the small Voles, is characteristic of damp environments. The Mole is another small creature but in this case feeds on the abundant worms and grubs of the rich grassland.

WORMS

EARTHWORM —
Lumbricus terrestris

Earthworms belong to a group of animals called the Annelids. This name comes from the Greek word for a ring since their bodies are made up of a series of similar ring-shaped segments. *Lumbricus terrestris*, which may measure up to 300 mm. long and have over 100 segments, is the largest European Earthworm, and is widely distributed over the Continent. Like other Earthworms it is pinkish in colour and has a number of tiny bristles on the lower part of each segment. These hold the position of the animal in the soil as it moves. Earthworms tunnel through the soil, eating out a passage for themselves and digesting humus and bacteria which may be in it. *Lumbricus* also comes to the surface at night and drags dead leaves down into its tunnel for food. All worms are hermaphrodite but not self-fertile; fully mature worms develop a clitellum or saddle — which is a ring of thickened tissue near to the front end of the animal. After mating with another of its kind, when sperms are passed to the other by both worms, the worm slides backwards out of the clitellum which now contains fertilised eggs. This forms a cocoon in the soil from which more worms soon hatch. Worms are long lived and may survive up to ten years. They are hugely numerous in meadow soils; up to 3,000,000 in half a hectare have been estimated.

SPIDERS

GLOSSY CROSS SPIDER —
Singa nitidula

This spider is inconspicuous because of its small size, for the female is only about 6 mm. long and the male less than this. It is, however, very widespread from France southwards into Spain and eastwards into Siberia. It also occurs in Cuba where it has been imported. If seen in tall grass or bushes in damp meadows, it may be recognised by its glossy, almost hairless abdomen, and its relatively short legs.

INSECTS

WARTBITER —
Decticus verrucivorus

Wartbiters are to be found through the warmer parts of Europe and temperate Asia, although they are extremely rare in Britain. They are bulky insects, measuring about 34 mm. long and with fairly long mottled wings, characteristic of open spaces with dense ground cover. In Britain they seem to be diurnal, and the males sing only when the temperature is high. In captivity they will accept plant food but in the wild will eat other insects as well. They get their curious name from the fact that they were used by Swedish peasants to bite off warts.

STRIPE-WINGED GRASSHOPPER —
Stenobothrus lineatus

Several colour variations are known of this Grasshopper, some of which are greener than the specimen illustrated. It is found in most of Europe other than the extreme north and south and is also known in Asia, usually emerging in early Spring and becoming adult by July. Its courtship song may be heard, especially in dry fields, from then until the Autumn. The eggs are laid at the roots of grasses and not in the soil as with most other Grasshoppers.

COMMON GREEN GRASSHOPPER —
Omocestes viridulus

This is a common species throughout Britain and

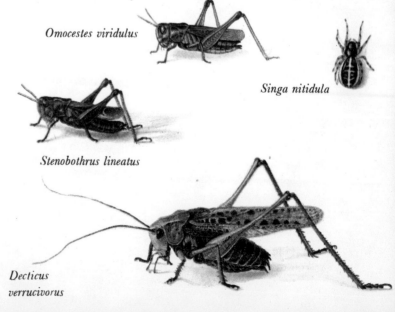

Omocestes viridulus

Singa nitidula

Stenobothrus lineatus

Decticus verrucivorus

Graphosoma lineatum   *Philaenus spumarius*   *Cercopis vulnerata*

Cercopis vulnerata

This Froghopper which is to be found in southern Britain and in Europe, is less abundant than the previous species, but is far more noticeable because of its bright colours. It may be seen in the summer months feeding on the leaves of willow or alder trees bordering damp meadows. The females lay their eggs in cracks of the bark where they overwinter, emerging next Spring to feed on the roots of grasses and other meadow plants. In their early life they are protected by a covering of froth.

Europe apart from the far south. Preferring lush grassland habitats, the nymphs hatch in springtime. Their song is continuous through the summer months. Common Green Grasshoppers, although 20 mm. long, are difficult to see and catch, and although they may be numerous they appear to be harmless.

SOLDIER BEETLE —
Cantharis fusca

This insect which measures about 12 mm. long, is one of the large group of Soldier Beetles, recognisable by their rather oblong shape and their soft wingcases. One of its near relatives is one of the most abundant and noticeable of British beetles in the late summer months; this species, while widespread in Britain, is not so frequently met with. On the Continent it is fairly generally distributed except for the far North. The larvae hunt other small insects and snails, mostly among the bases of the stems of plants during the daytime, but ascending the plants at night to capture other prey. The adults feed on insects and some plant material. They overwinter in the larval stage, pupating early in the Spring and emerging about midsummer.

Graphosoma lineatum

This brightly coloured bug, which measures about 10 mm. in length, is a common summer inhabitant of the southern parts of Europe, where its bright colours make it an obvious inhabitant of meadows and fields, usually close to streams. In some areas it may cause damage to crops such as carrots, but in spite of its widespread presence and the fact that it tends to occur in groups, it is avoided by insectivorous animals in general because of its unpleasant flavour.

FROGHOPPER —
Philaenus spumarius

Over much of Europe the froth of Cuckoo spit on thistles and other plants is a familiar sight in meadows in early Summer. This is the hiding place of the nymph of the common Froghopper which is a soft bodied creature, which one would expect to be vulnerable to the attacks of many predators. Its protective cover is formed from a liquid secreted from the anus and abdominal glands and made into a foam by air bubbles forced into it from the underside of the body. At least one insect predator can penetrate the disguise; this is a solitary wasp which drags the helpless nymphs from their hiding place and provisions its larval cells with them. Most Froghoppers seem to survive, nonetheless for later in the summer the adults are normally very abundant. These are dull coloured insects about 8 mm. long and capable, as their name suggests, of leaping several metres at a time if disturbed, or of flying if need be.

Malachius aenus

This is another beetle with soft wingcases, which may sometimes be seen on meadow flowers in early Summer, for the adult feeds on pollen. The larvae are predatory, feeding on grubs in old wood and sometimes entering the tunnels of Bark Beetles to hunt the occupants. It is not very common in Britain, and is even rarer towards the north. In Europe it is widespread and it also occurs in Canada and around New York.

Sisyphus schaefferi

In classical mythology Sisyphus was a man condemned for ever to roll a huge stone uphill. This 10 mm. long beetle has been given the same name because it may be seen in meadows of the warmer parts of Europe and near Asia, trundling a relatively huge ball of dung along. Usually a pair of insects is involved, rolling the ball which they have formed backwards, using their hind legs to propel and guide it. Eventually, they find a place where they can excavate a chamber and bury their burden. The female lays her eggs in it and it serves as food for the larvae. In the course of their lives many such dung balls will be made and the adults will themselves feed on it.

DUNG BEETLE —
Aphodius distinctus

The big herbivores digest their tough food relatively incompletely, and their droppings, therefore, contain nourishment for a large number of dung eaters. This little beetle, which measures about 5 mm. long, may be found in cow pats in Britain, Europe and North America, protected from the clinging qualities of its environment by its extremely waxy coat. Dung Beetles are most valuable members of the fauna, in that they return to the soil minerals from the dung.

TORTOISE BEETLE —
Cassida viridis

There are many species of Tortoise Beetle, some very

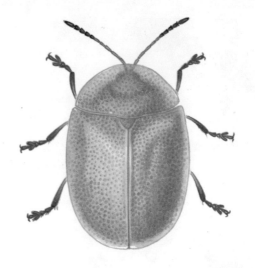

Cassida viridis

Cantharis fusca

Sisyphus schaefferi

Malachius aenus

Aphodius distinctus

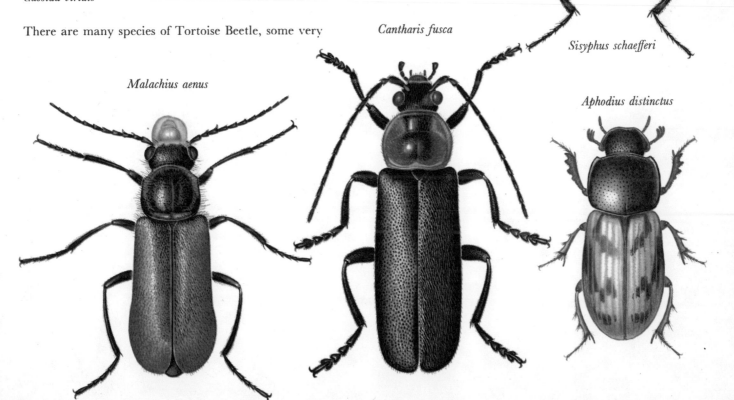

Panaxia dominula

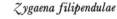

Zygaena filipendulae

Euplagia quadripunctaria

♂

Melanargia galathea

♀

Heodes virgaureae

colonies so several may often be seen at once, crawling over the same stem or flower head. Their extremely unpleasant taste protects them from predators which are reminded of this fact by the moth's vivid warning colours. In Britain this species is generally found on the warm dry slopes of chalk or limestone hills or near the coast and it is widespread in Europe. The moth belongs to a family which has clubbed or toothed antennae. The caterpillars are sluggish yellow-green creatures with black spots; they feed on the leaves of various members of the vetch family. They hibernate before being fully grown and complete their development early the next Spring. The cocoon of the Six-spot Burnet Moth is a silvery parchment-like object attached to grass stems.

SCARLET TIGER MOTH —
Panaxia dominula

Scarlet Tiger Moths have a wingspan of about 50 mm., but in spite of their gaudily coloured hindwings they are not normally seen for they are active at night, hiding by day on the bark of trees bordering damp meadows. In Britain it occurs mainly in the southern counties and becomes extremely rare further north, but it is widespread in mainland Europe. The caterpillars hatch from their eggs in July or August and feed on a wide rang of meadow plants, including nettles, groundsel, comfrey and brambles, and sloe and sallow among the hedgerow plants. They hibernate early and complete their growth next Spring, finally emerging as an adult moth in June. There are many colour varieties of this moth to be found particularly in Southern Europe and Western Asia.

JERSEY TIGER —
Euplagia quadripunctaria

This handsome moth, which has a wingspan of about 60 mm., is common throughout Southern Europe, occurring as far north as the Channel Islands. During the last century it began to spread north into Britain although it has never become common there. The caterpillars feed on various plants typical of damp hedgerows or meadows, and sometimes on elm leaves. They complete their growth after hibernation then

similar in appearance to the one figured here, which is 7 to 10 mm. long and usually found on plants of the mint family in damp meadows. It occurs in Britain, although it is rare in the north, and across Europe and Asia to Japan. Both the adults and the larvae feed on plants. The larvae are somewhat flattened and have a row of blunt spines along the sides. It is protected further from a covering of its own excreta which becomes attached to a projection at the rear end of the body.

SIX-SPOT BURNET —
Zygaena filipendula

The Burnets are unusual among moths because they are active by day rather than night. They are not very large (35 mm. wingspan) but are brightly coloured and generally slow moving. They live in spin a flimsy cocoon at the base of the plants, emerging as adults in July.

MARBLED WHITE —
Melanargia galathea

The black markings on the white wings of this butterfly caused it to be given the old English name of the 'Half Mourner'. It is found in Central and Southern Europe and east into Asia minor. In Britain it occurs from the Midlands southwards, but is most abundant on the chalk downlands where it may occur in small colonies. The caterpillars feed on grasses of various kinds, hibernating when they are very small and completing their growth during the next Summer. After a brief pupation period, the 45 mm. wingspan adult emerges.

QUEEN OF SPAIN FRITILLARY —
Issoria lathonia

This elegant little butterfly, with a wingspan of about 45 mm., is characterised by the large silvery spots on the underside of the hindwings. It is an inhabitant of North Africa and Southern Europe from where it migrates further north. It occasionally reaches Britain but is not a resident. Its activity is dictated by local conditions, for in areas of warm climate the first summer generation is active in February and two or three broods, which feed on violets, may be produced in a year. In colder mountainous or more

northerly areas only one brood is produced and it is said to hibernate in any state of development.

SCARCE COPPER —
Heodes virgaureae

The gleaming orange-gold of the male and the heavily black spotted female of the Scarce Copper Butterfly may be seen in Central and Northern Europe, feeding on the flowers of golden rod in July and August. A number of forms occur, including a smaller sub-species, with a wingspan of only about 30 mm., which is found on mountain slopes above about 2,000 metres. The larval food plants are various species of dock.

LARGE BLUE —
Maculinea arion

This butterfly occurs over much of Central and Western Europe but in Scandinavia only in the south and in small areas of central Spain. In Britain it is now extremely rare and lives in a few places in southern England.
The adults are active in the midsummer months, when the eggs are laid on plants of wild thyme. This is the early food of the caterpillars but after their third moult they lose interest in plant food and wander about until they are discovered by an ant. The ant takes the caterpillar off to its nest, and from

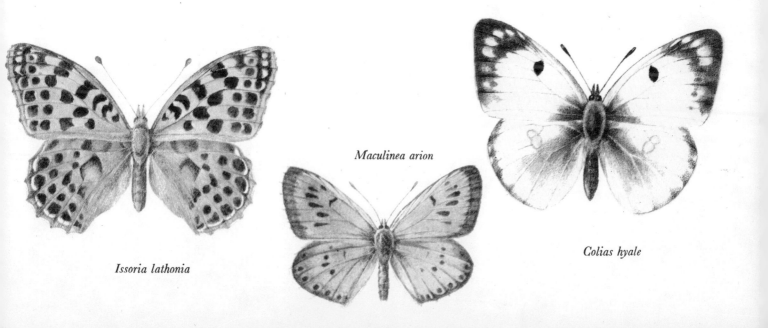

Issoria lathonia

Maculinea arion

Colias hyale

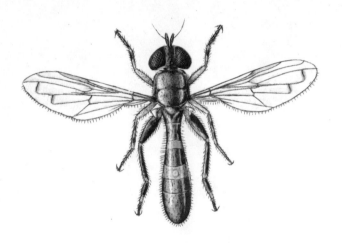

Neoascia podagrica

then on it feeds on the young larvae of its hosts, a habit which is tolerated by the ants for the sake of honeydew which it produces and which they crave. It overwinters in the ant's nest where it pupates unmolested. The adult butterfly, however is able to escape undamaged from the tunnels of the ants after it emerges, for the expansion of the 25 mm. wingspan wings are delayed for some time.

PALE CLOUDED YELLOW BUTTERFLY —
Colias hyale

This insect may occasionally be seen in southern England, for although it has a wingspan of only 45 mm., it is strongly migratory and may fly long distances from its mainland European home. Two distinct broods are found, one in early, the other in late Summer, in flowery meadows where the eggs are laid on plants of the vetch family which are the food of the caterpillars.

CRANE FLY —
Tipula paludosa

This species of Crane Fly is the most common of its kind in arable land in Britain. The adults, which are some-times called by the descriptive name of Daddy-long-legs, have drab grey bodies, and are to be seen from midsummer onwards when the females lay their eggs in meadows and fields. These hatch into grey grubs known as leatherjackets, which are often very numerous and can become a major pest, for they feed on the roots of grasses and other plants and fields may show large bare patches where they have been at work.

Neoascia podagrica

This little hoverfly which measures only 5—6 mm. in length, is widespread in Britain and Europe and is common in damp meadows flying especially around clumps of nettles. It may also occur in gardens where it hovers round herbaceous plants, or may be seen resting on fences or gateposts. Although small, it seems that the males may be territorial in their behaviour, defending suitable breeding places against others of their own kind.

BIRDS

MEADOW PIPIT —
Anthus pratensis

The Meadow Pipit is to be found as a resident in Britain and adjacent parts of the Continent. In Scandinavia and North East Europe it is a summer migrant, departing after breeding, for Southern Europe, Western Asia and North Africa where it spends the winter months. In general appearance the Meadow Pipit resembles the Skylark but is slightly smaller (145 mm. long), slimmer, and streaked with brown above and below.

Outside the breeding season, when about four or five young are reared in a nest made on the ground, Meadow Pipits associate in loose flocks, searching for the seeds, insects and spiders on which they feed. On the ground they walk and run rather than hop. If frightened they take to the air with thin calls of alarm. The song of the male, unlike the rich music of the Skylark, is a thin trill usually given in a short song-flight with a 'parachuting descent'.

188

YELLOW WAGTAIL —
Motacilla flava

The Yellow Wagtail is a summer visitor to almost the whole of Europe, but it is extremely variable in colour and a large number of subspecies, based on plumage differences, have been described. The bird illustrated is the Blueheaded Wagtail, which is the common mainland European form. This does not occur in Britain, where another race, with a yellower head and less well marked dark patch behind the eye, is to be found, although this latter species does not reach Scotland or Ireland. All races live in well watered open areas where they feed on worms, molluscs, spiders and insects which they may catch on the wing. They often associate with grazing cattle and catch the small creatures which these disturb. The nest is built in thick cover on the ground and here one or sometimes two broods of young are reared. After the breeding season the Yellow Wagtails gather into flocks, usually living near to large stretches of water, before migrating to their winter quarters in Central Africa.

WHINCHAT —
Saxicola rubetra

Whinchats are rather stockily built birds of open country, found in Britain mainly on heaths and commons where bushes and grass grow together, although they may occur in cultivated fields especially when migrating. About 125 mm. long, the Whinchat lives over almost all of Europe except for the far north and the Mediterranean countries. The noises made include a number of churring and clicking sounds but the song is a brief musical warble. The nest is made in the shelter of a bush or dense clump of grass, and the young reared on the same small invertebrates as the adults feed on. In wintertime Whinchats migrate to Africa.

STONECHAT —
Saxicola torquata

The Stonechat, which is a close relative of the Whin-

Upupa epops

Motacilla flava

Saxicola rubetra

Ciconia ciconia

chat, is similar in size (125 mm. long) but of even stockier and rounder appearance. It is easily distinguished by its colours, for the male Stonechat has a black head and neck with a pronounced white patch below the ear coverts. The back and tail are black but the rump is pale grey; the wings are also dark with a white wing bar. The underparts are a similar colour to those of the Whinchat, but are slightly deeper in shade. The female is less intensely coloured, with a dark speckled back and wings and her dark head distinguishes her from a male Whinchat for which she might otherwise be mistaken. Both species inhabit similar country in Western and Southern Europe where the Stonechat is resident. In Central and Eastern Europe it is a summer migrant, and except in Britain is not found north of the latitude of Holland. It often sits in exposed places, jerking its wings and tail, and the male may sing his rather simple song in flight.

HOOPOE —
Upupa epops

The Hoopoe's curious name is onomatopoeic and imitates the deep crooning noise made by this bird. There are a number of breeding records for this species in Britain where it is a rare summer visitor. On the Continent it is widespread, as far north as the Gulf of Finland, although it migrates to Africa in the winter months. Hoopoes, which measure about 280 mm. long, are quite unmistakeable as they walk and feed in open country or parkland with scattered trees. The large crest, shown erect in the illustration, can

be folded back to give the bird an almost anvil shaped head and in their undulating flight the rounded wings show a dramatic pattern of black and white stripes. They feed on grubs and worms which they pull from the ground with their long bills, probing deeply in the rich meadow soil for creatures not available to birds with shorter beaks. Hoopoes nest in holes in trees or in old masonry. No nest material is used and the female begins incubation as soon as the first egg is laid with the result that the brood hatches on different days and until fledged the young are of different sizes. The male feeds the female during incubation and later both share the duties for caring for the young. Hoopoes are among the very few birds which do not practise nest sanitation so the nursery is a foul place. It was perhaps for this reason that Hoopoes, along with their relatives the Kingfishers, were thought to be unclean animals and therefore were declared unfit food for the children of Israel.

SHORT EARED OWL —
Asio flammeus

This owl is resident in Britain northwards from the Midlands and across the Continent from France to Western Asia. It breeds in Scandinavia and latitudes east of this but retreats from these very cold areas to southern parts of the Continent in wintertime. The Short Eared Owl which measures about 375 mm. in length is more active in daylight than most other owls. It may be seen quartering the meadows with a buoyant, gliding flight with its long wings tilted into a V-shape like that of a Harrier; but it may be distinguished from a hawk by its short tail and large head. In springtime the male's display includes a high circling flight with sharp cracking noises produced by smacking the wrists of the wings above and below the body, and it is at this time that the ear tufts come into use, for they are in no way concerned with hearing. The Short Eared Owl feeds very largely on Field Voles and other small rodents which have population fluctuations leading to plague numbers in some years. The owls gather in some areas where such food is plentiful and benefit to the extent that they may be able to rear twelve or fourteen young rather than the usual four or five.

190

HEN HARRIER —
Circus cyaneus

The Hen Harrier may be present as a nesting bird through much of mainland Europe, except for the extreme south; although after breeding it migrates from the coldest areas to as far south as North Africa. In Britain it is resident only in Scotland, but may be a winter visitor elsewhere. It is smaller than the Marsh Harrier (see page 98), for its wingspan does not exceed 1,100 mm. nor its length 510 mm., but in general methods of flight and hunting the two species are similar, gliding low over the ground searching for small prey such as Voles. The male can be distinguished easily by his pale grey and white colouring, although the female is brown. The nest is built on the ground often in marshy meadows, and the female, as with all hawks, begins incubation as soon as the first egg is laid. The male feeds her and later, when the chicks have hatched, after about 30 days incubation, he provides for all the family, often plucking the food before delivering it to his mate. The female soon joins him in his hunting to keep pace with their appetites.

WHITE STORK —
Ciconia ciconia

Among the larger European birds the White Stork is the best known and the most numerous, apart from the semi-wild Mute Swan. This bird, which has a length of 1,020 mm. and a wingspan of 1,680 mm. is a relative of the herons and like them flies with its neck stretched out and its legs trailing. At one time it nested in trees on cliffs throughout much of continental Europe. Today it usually nests near towns or buildings, sometimes on them, and as it is traditionally regarded as a bird of good omen, it is not molested but encouraged, in many cases with specially built platforms for its use. However, beyond Central and Eastern Europe the White Stork has become relatively rare for there is no such regard or protection for it. Unlike most birds the White Stork is mute apart from some grunting and hissing noises, but clatters its bill as part of its springtime display. A pair of birds will return for many years to their nest site, renovating it each Spring, and rearing two to five young there. The male helps with incubation which lasts for about 30 days, taking his turn during the daytime while the female sits on the eggs at night. Storks forage in open farmland and meadows for the insects and other animals including fishes and frogs, snakes and small mammals which are their food. The nestlings are fed at first on partly digested food, but later are brought complete small creatures. It is about 40 days before they are fully fledged, but before they leave the nest they practise wing-flapping to strengthen the muscles which they will need for flight. Storks migrate to South Africa, often in huge flocks. Although the fact of their migration has been known since biblical times, details of it have only become clear recently. The results of large-scale ringing have shown that populations from Eastern and Western Europe migrate down the opposite sides of the Mediterranean and Africa.

CORNCRAKE —
Crex crex

Corncrakes are summer migrants to much of Europe north of Spain and Italy, but in Britain at least are one of the species which is becoming increasingly rare for natural causes, rather than because of human interference. It is a secretive bird, about 270 mm. long, belonging to the Rail family, mottled brown in colour although the males have a grey throat patch and both in flight show the red colour of their secondary and shoulder feathers. Although rarely seen, the Corncrake has a call which makes its presence obvious during the breeding season, for the monotonous harsh 'crex-crex' call from which its latin name is derived may be kept up all night. The nest is well hidden in tall vegetation, and the 7—12 eggs which are incubated for 21 days hatch out as completely black chicks. Corncrakes feed on both plants and animals, including many insects and spiders.

191

IN
WARM
PLACES

INTRODUCTION

Europe is alone among the inhabited continents in being entirely outside the tropics. Nonetheless, in the south, summer temperatures may be considerable and in the countries bordering the Mediterranean, snow and frost are rare, even in wintertime. The plants of these areas may have to be able to withstand prolonged summer drought. We find that many are adapted to resist water loss with small, leathery leaves and many contain aromatic oils which give the southern flora a characteristic scent, such as that of the lavender or thyme which clothes the hillsides.

Although Europe no longer has elephants, hippopotamuses or rhinoceroses as native creatures, some of the animals of the Mediterranean region are closely related to, or even the same as the African fauna, for here we find porcupines, genets and termites, to name only three. The proverb 'Africa stops at the Pyrenees' seems apt when we look at the hot, dry lands of Spain and parts of Italy and the Balkans. It is not only in the south that we find conditions of warmth in Europe, although elsewhere they are likely to be short-lived and replaced in Winter by biting frosts and freezing winds which make it impossible for any large warmth-loving creature to survive. There are, however, many small animals, such as some insects, which can spend the cold part of the year in hibernation, well hidden against the worst that the weather can offer, deep in the soil, or in cracks in rocks. In mountainous areas the rock faces, even in the north, may get very warm during the Summer, but are bitterly cold in the Winter. Here, as in the south, we find that the colonising plants often need to retain water and the thick, fleshy leaves of stonecrops or houseleeks which find a roothold in the most precipitous places on cliffs, or the man-made environment of roof tops slopes, contain liquids to sustain them in the dry summer months. Many small insects are to be found in such places preyed on by spiders which are not, on the whole, builders of complex orb or hammock webs, but rather hunting and jumping species, which pounce on their prey having sighted it from some distance off.

In Central and Eastern Europe the summer temperatures are very high and again we may meet conditions suitable for warmth-loving creatures. In the lowlands, however, most of these have disappeared, for man's takeover of the environment has meant sweeping away many native species of plants and animals, although in a few areas, kept as nature reserves, the flora and fauna of the past is preserved. Sand dunes are another habitat in which very high temperatures may build up. Unprotected for the most part by vegetation, the dune surface may become very hot. Uninhabited as they may appear, they nonetheless house many creatures, such as the Sand Wasp and a whole web of predators and prey. Many of these are active only at night, when the temperature is lower and dew may give some vital moisture, but early in the morning we may have an inkling of their presence, from the maze of curious tracks, which criss-cross the sandy slopes.

Britain is too far north, too cold and too wet, especially in the Winter, for the survival of many hot climate species. The two habitats which come closest to conditions of real warmth and dryness are the Chalk Downs and in some places southward facing slopes in road or railway cuttings.

The Downs, made of porous chalk, are dry, and here one finds many species of plant and a few animals which have their strongholds in the south. Some of the wild orchids for which the Downs are famous, such as the Bee and the Fly Orchid, which are comparatively rare and found only on the chalk, are abundant in Southern Europe and grow on a wide variety of rock types. Several species of insects, some characteristic of warmer climates, are also found there. Road, or more especially railway cuttings may offer on their south facing slopes small areas of specially favourable climate, in which limited populations of warmth-loving plants and animals may survive.

SPIDERS

ZEBRA SPIDER —
Salticus scenicus

This little jumping spider is found widely through Britain and Europe where it is often seen on sunny walls and fences. Although it protects itself with a silk line which it attaches to the surface as it goes along, it does not make a web or snare to trap the small flies and other insects which are its prey. Instead, it stalks them and pounces, often from some distance away to catch them. Unlike most other spiders, it has excellent vision and its large eyes may be seen on the front of its head. During the breeding season the male, which can be distinguished by his enormous fangs, approaches the female with a sort of semaphoring dance, by which she recognises his presence as a suitor, rather than a meal.

Eresus niger

The name of this spider refers to the female, which is about 16 mm. long and completely black in colour, while her mate, which is only half her size, is coloured as shown in the plate. It is to be found in the warmer parts of Europe and has often been recorded from the extreme south of England, although it has not been seen there for many years. This spider is one that lies in wait for its prey in a burrow, up to 10 cm. deep, which is dug in sandy soil and supported and strengthened with a silk lining. A lacey silk cover extends over the mouth of the hole and if a beetle or other large insect walks over it, the legs will sink through, and the spider will bite one of them. In spite of her size, the spider will wait until her prey is completely paralysed before pulling it through into the burrow so she can feed.

INSECTS

PRAYING MANTIS —
Mantis religiosa

The Praying Mantis, which is found in the warmer parts of Southern and Central Europe, gets its name from the position, reminiscent of hands folded in prayer, in which it normally holds its front legs. This pious attitude is deceptive, for the Praying Mantis is, throughout its life, a predator, waiting for a victim. When a suitable small insect, such as a fly or beetle appears, the folded forelimbs will be shot out to grasp the creature in a hold from which there will be no escape. Camouflaged by its green colouring, the Mantis does not need to be an active animal; mistaking it for a part of the herbage, many insects will approach without realising their mistake until it is too late. The size of the Praying Mantis varies greatly, according mainly to the avialability of food, but the female may reach a length of 75 mm., while the more elegant male grows to a maximum of 60 mm. Courtship and mating is fatal for the male, for early in the proceedings the female seizes him and begins to eat him, starting with his head. Gruesome as it may seem, the severance of the male's main nervous ganglia is a stimulus to his mating behaviour. After mating and her meal are finished, the female lays her eggs on twigs, protected in a case of hard, foam-like material.

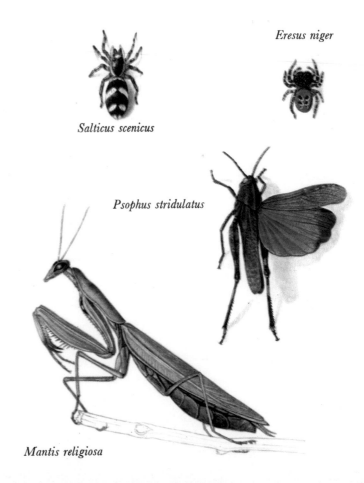

Eresus niger

Salticus scenicus

Psophus stridulatus

Mantis religiosa

RED — WINGED GRASSHOPPER —
Psophus stridulatus

This species is found in areas of Europe and Asia east to Mongolia, where the Summers are warm. Even if the Winters are cold, this does not affect the insect, which passes this time of year in the security of its egg. In flight, the adult shows bright red hindwings, which deceive predators into looking for a red object, even when the insect has landed and is camouflaged by the green colour of the rest of its body.

BLUE — WINGED GRASSHOPPER —
Oedipoda coerulescens

This grasshopper is widespread across Europe, Western Asia and North Africa. It occurs in the Channel Islands, but is not found in mainland Britain. Like the previous species it feeds on plants and escapes from enemies by flight, when it shows the vivid blue colour of its hind wings.

RUBY TAILED WASP —
Chrysis ignita

Ruby Tailed Wasps are abundant in the warmer parts of Europe. However, they are rarely seen, despite their bright colours, for they are small, measuring only about 8 mm. in length and although the adults spend part of their time feeding on nectar from meadow flowers, much of it is passed looking for suitable places to lay their eggs. While doing this they are unlikely to be noticed, for they search the ground looking for the nests of solitary wasps. Having found one, the *Chrysis* female lays an egg on the larva of the wasp, although if, while she is exploring the nest, the rightful owner returns, she suffers no harm. She merely rolls up into a hard-cased ball, which suffers no damage when thrown out by the wasp. When the *Chrysis* egg hatches, the larva feeds by sucking the body fluids of the living grub of its host.

VELVET ANT —
Mutilla europaea

In spite of its name, the Velvet Ant is really a kind of wasp, but because the female is wingless, she has an ant-like appearance. It is found mainly in Southern and Eastern Europe and occurs also in the southern counties of Britain, although it is never common there. Like the previous species the Velvet Ant is a parasite, but in this instance the female seeks out the nests of Bumble Bees, the larvae of which are the hosts on which her grubs feed.

Anoplius iraticus

This solitary wasp is common on dry sandy heaths and similar warm places in Britain and Europe. Here the female digs out a burrow in the sand and provisions it with food for her young. Like all solitary wasps she will take prey of one kind only and in this instance specialises in spiders. Orb-web builders are ignored, but the free-running Lycosids, which are abundant in heathlands are attacked. Having found a suitable spider, she paralyses it with her sting and takes it back to her nest, and once it is safely underground, she lays an egg on it. This hatches into a grub which feeds on the spider, which although helpless, is not dead. Several spiders and eggs may be placed in a single burrow. When fully fed, the grubs pupate and emerge during the next Summer.

SANDWASP —
Amophila sabulosa

The Sandwasp, which is 18—20 mm. long is an inhabitant of dry heaths and sand dunes in Britain and Europe. Like the previous species it is a solitary wasp, but hunts caterpillars rather than spiders. Unlike some of her relatives which dig their nest hole leaving a tell-tale 'tip' outside, *Amophila* carries the sand away and scatters it, so that the entrance to the hole is scarcely visible. In doing this, she learns the position of the nest in relation to its surroundings, but before leaving it to go hunting she closes the entrance with a pebble or piece of wood, so that it will not be seen by any enemies. Having found a caterpillar, she stings and paralyses it, then takes it back to the nest. Since the prey is a large creature in relation to the predator, it has to be dragged over the ground and the Sandwasp often has to verify its position by climbing up plants to check the landmarks by which it navigates. Each nest contains only one egg, but the female returns to provision the larva with further caterpillars as it grows, apparently remembering the positions of several nests in which she may have developing grubs at any one time.

Andrena carbonaria

This solitary bee, which occurs widely in Britain and Europe, is active mainly in the springtime. It is about the size of a honey bee, but is easily recognisable by its almost entirely black coloration. Like the wasps already described, each female digs a tunnel in sandy soil, but makes provision for the young with pollen, which is collected from many sorts of flowers. The adult is restricted in its food supply, for its tongue is short and it is forced to feed from flowers in which the nectaries are not deeply hidden. Although each female provides for only her own young, groups of burrows may be made close together, giving the appearance of a colony.

Eucera longicornis

This insect, may be recognised by its bee-like appearance, coupled with long antennae, which are especially well developed in the males. It is common in Southern Europe and in warm places in central and eastern parts of the Continent. It occurs in sheltered habitats in southern Britain, where as elsewhere its long tongue enables it to feed on nectar from many sorts of meadow flowers.

DUNE TIGER BEETLE
Cicendela hybrida

This Tiger Beetle is darker in colour than the other kinds shown on pages 20 and 32, and measures only 12—14 mm. long. Although abundant in Southern Europe, it is an extreme rarity in Britain and may be found only in a few places in the south. Its general way of life is like that of the other Tiger Beetles: an extremely active ground predator, it feeds on any small animal which it can subdue. The larva excavates a tunnel and lies in wait near the top of it, supported

by a hump on its back. It will grab and devour any creature which approaches.

SPANISH FLY —
Lytta vesicatoria

In spite of its name, the Spanish Fly is really a beetle, which is common in Southern Europe but rare further north although a single colony is reputed to exist in Britain. Its beautiful metallic green colour camouflages it against the background of the leaves of ash

Chrysis ignita

Oedipoda coerulescens

Mutilla europaea

Anoplius iraticus

Amophila sabulosa

Lytta vesicatoria

Eucera longicornis

Myrmeleon formicarius

Ascalaphus macaronius

Chazara briseis

Parnassius mnemosyne

Zerynthia polyxena

trees, on which it feeds, but it has a further protection, against predators in a substance called cantharadin, which it produces. This is a powerful blistering agent and causes a painful rash to anyone who touches the insect. Its use as a supposed aphrodisiac has made this beetle better known than most of its relatives, but cantharadin is a potent poison and anyone using it courts danger. The female lays her eggs in grassy places and the larvae lie in wait for solitary bees to which they cling. In this way they get carried back to the nest, where they become kleptoparasites, feeding on the food supplies intended for the bee's brood.

Ascalaphus macaronius

At first sight this insect, which is found in Southern Europe, might be mistaken for a rather slow flying butterfly. A closer look, however, will show that it has gauzy, net-veined wings of similar size, with the 'laddered' leading edge characteristic of the Order Neuroptera. Like its British relatives, the Lacewing Flies, when at rest *Ascalaphus* folds its wings to make an inverted V-roof over its back. Both the adult and the larva are carnivorous, the latter living on the ground and hunting ants in preference to other prey. It does not make a pit like its relative the Ant Lion, but is a successful hunter all the same, sucking its captives dry with its large mandibles.

ANT LION —
Myrmeleon formicarius

The adult Ant Lion, which is figured here, is probably not a very familiar insect to most people, for it has a brief life and may often be mistaken for a small dragonfly, with a wingspan of about 75 mm. Its method of folding the wings when at rest and the prominent stout antennae should show rapidly that it is a Neuropteran. The larval Ant Lion is much better known, however. In dry, sandy areas of Southern Europe it digs a shallow pit and hides in the loose material at the bottom. Any little creature, such as an ant, which steps on to the sloping sides of the hole starts the sand grains rolling and cannot keep its footing to climb out again. The larva encourages the movement of the slope by throwing sand grains up from the bottom of the pit. The ant slides rapidly down and is seized by the huge mandibles of the waiting grub. Having pierced the outer covering of the ant, digestive juices are pumped into it, down deep grooves in the mandibles. Once it is converted into a liquid by this means, the larva then sucks up its predigested meal.

HERMIT BUTTERFLY —
Chazara briseis

This butterfly is found in Central and Southern Europe, North Africa and Western Asia, usually in dry, stony places, occuring up to a height of 2,000 metres in the warmer parts of its range. The pale patches on the hindwings are usually more marked than those shown in this illustration. Other variations include size, for in the south the male may be up to 60 mm. in wingspan, while further north it may be as little as 45 mm. The caterpillar, which is yellowish grey in colour, with a dark stripe down its back and on its side, may be seen in the Autumn, feeding on grasses. It overwinters in the larval stage, and continues its development next year, pupating in June and emerging shortly afterwards.

CLOUDED APOLLO —
Parnassius mnemosyne

The Clouded Apollo is widely distributed in Central and Southern Europe and across the Caucasus to Central Asia, where it may be found up to a height of 1,500 metres, although further north it is a lowland species. It has a wingspan of up to 60 mm. and is active in May and June, when it is often seen flying at dusk, rather than in bright sunlight. The caterpillar, which is red, with rows of black spots, feeds on fumitory.

SOUTHERN FESTOON —
Zerynthia polyxena

The Southern Festoon is a butterfly of Western Asia, and is also known from South East Europe, although it is nowhere common. It is on the wing in April and May, usually in rough, stony places, although it

is rarely found above an altitude of 1,000 metres. The caterpillar, which feeds on various species of brithwort *(Aristolochia)*, is reddish-brown in colour, with pale red projections on the body.

REPTILES

GREEN LIZARD —
Lacerta viridis

The Green Lizard, which may grow to a length of 500 mm., inhabits Central and Southern Europe, reaching as far east as south-west Russia and as far north as the Channel Islands. Several attempts have been made to introduce it into Britain, but none of these has been successful, although the animals have survived for several years in some cases. Apart from its colour, the Green Lizard may be recognised by its very long tail, which is at least twice the length of the head and body combined. It may, however, be broken especially in fighting males in the springtime. It is usually found in dry areas where there are sunny banks, but with vegetation or stones under which it can hide,

Lacerta viridis

Lacerta muralis

Ablepharus kitaibelii

for it is very shy and disappears at the least sign of danger. It is extremely agile and can climb quite steep banks or rocks where it may rest and sun itself. It feeds on insects, spiders, worms and grubs of all kinds and in captivity will also eat soft fruit, though this is not likely to play a great part in its diet in the wild. In the summertime the female lays a small number of eggs (usually not more than 20 and sometimes as few as 5) which she buries in the ground, leaving them to be hatched out in the warmth of the sun. The young hatch in 6—8 weeks and have grown several centimetres by the end of the Summer. When, as the cold weather approaches, they find secure, frost free places in which to pass the Winter in a state of hibernation.

WALL LIZARD —
Lacerta muralis

The Wall Lizard is found from Holland southwards and across Central and Eastern Europe, but prefers warm, dry, rocky country, where it basks in the sun. It is like the previous species in shape, with a long, tapering tail, but is much smaller, the males reaching a length of less than 200 mm., while the females are still smaller. The basic colour is between brown and grey, but there is a large number of geographical varieties, which may be green or black. The Wall Lizard is well named, for it has long, sharp claws which enable it to grip and climb walls, rocks or trees where it can find insects, spiders or grubs on which it feeds. In May and June the females lay several batches of eggs, between 2 and 8 in number, which they bury in the ground to hatch in the heat of the sun.

Ablepharus kitaibelii

The little lizard, which grows to a length of about 110 mm. is a member of the Skink family, which is a group well represented in the tropics, but with few European species. *Ablepharus*, like all members of the group has a rounded body with small, smooth scales and tiny legs. It is found in the extreme south-east of the Continent and the warmer parts of Central Europe, where it inhabits rocky areas on the edge of woodlands and remote farmsteads. It feeds mainly

on insects and in the summertime the females lay eggs.

BIRDS

BLUE ROCK THRUSH —
Monticola solitarius

This bird, which measures 200 mm. in length, gets its name from the bright blue-grey colour of the body of the male. His tail and wings are black and the female is dark brown barred and speckled below with pale buff. In Europe it is to be found as a resident in Spain, Italy and part of the Balkans. Although the musical, thrush-like song of the males may often be heard, the birds are shy and dive for cover among the boulders when approached. They feed on quite small invertebrates and berries.

WHEATEAR —
Oenanthe oenanthe

The Wheatear is a summer migrant to the whole of Europe, making for open country, where it can nest in cavities in rocks or walls, or in disused rabbit burrows. The bird figured is a male; the female is brown above but has the same pattern of black on her tail and the white rump which in both sexes is conspicuous in flight. Both birds are about 145 mm. long. The song of the male is a brief, musical warbling but both sexes make a monotonous clacking alarm call, which is one of the most characteristic sounds of moorlands and wild upland country. Wheatears feed on insects, grubs and spiders.

BEE EATER —
Merops apiaster

The 280 mm. long Bee Eater is a summer migrant to Southern Europe, preferring rather open country with few trees. Occasionally, in a warm year, it may reach as far north as Britain, where it is known to have bred successfully in one season this century. It is related to the Kingfishers and Hoopoe and like them, nests in holes which it digs to a depth of more than

Oenanthe oenanthe

Merops apiaster

Neophron percnopterus

one metre in sandy banks of rivers or in gravel pits. Since suitable sites are often in short supply, Bee Eaters usually nest in large colonies. As its name suggests, it is an insectivore, and may be seen sitting on telegraph wires or posts, scanning the area for sui- table food, or hawking after insects with its graceful, swallow-like flight, sometimes even hovering. Its prey includes wild and honey bees, butterflies and dragon-flies.

Genetta genetta

a kilometre away, which in turn alerts others to the possibility of food. The result is that a number may arrive quite quickly to act as a demolition squad on the remains of any dead animal, and may be valuable removers of unwanted carcasses in the areas where they occur. Vultures' nests are bulky affairs, made of sticks, placed on a fairly rocky ledge protected by an overhang. The normal clutch is two eggs, which are incubated for 40 days by both parents, who then have to care for the young for a long time before they are independent. At first the fledgelings are dark brown in colour, but they become paler as they mature and achieve their white adult plumage when they are six years old.

MAMMALS

GENET —
Genetta genetta

The Genet, which at first sight looks like a member of the weasel family, is in fact different in many fundamental anatomical details and is more closely related to the Mongooses and Civets of the tropics. It has a long body (470—580 mm.) and an almost equally long tail, but its height at the shoulder is only about 200 mm. Although it occurs throughout Spain and much of France and occasionally turns up in Belgium, Holland and Germany, it is nowhere common. It is rarely seen because of its nocturnal habits and shy behaviour, although it may sometimes be detected by a heavy musky odour, which is the product of the anal glands. This, like the gland of the Civet, used to be used in perfumery and in various folk medicines. The Genet normally lives in thick cover in woods where streams have cut rocky valleys and where it can make a den in a tree stump or under a boulder. It can climb, leap and swim well and feeds on a wide variety of invertebrates and small vertebrates. Although they occasionally venture into farmsteads and take a hen, Genets are generally tolerated by man because of the large number of rodent pests which they kill. Three young are normally produced each year. These are born in a secure den, where they are cared for by their mothers for several months.

EGYPTIAN VULTURE —
Neophron percnopterus

Several species of vulture are to be found in Southern Europe during the summer months, but by far the most numerous is the Egyptian Vulture, which may be seen over Spain, southern France, Italy and the Balkans as well as in Western Asia and in Africa. It has been recorded as far north as Britain, but it only has the status there of an extremely rare vagrant and certainly never stays to breed. It is the smallest of the Vultures occurring in Europe, with a length of up to 660 mm. and a wingspan of 1,500 mm. It may be recognised in flight by the broad 'fingered' wings, the small head and the wedge-shaped tail. On the ground its large size, pale colour, rather flat, large feet and bare head with a relatively straight beak are recognition points. The Egyptian Vulture is a scavenger, regarding no animal food as too putrid to be eaten. It often haunts rubbish dumps, where it may find something to its taste, but may also fly long distances, circling high on ascending currents of warm air, before it can spot anything edible. When it does so, it descends at once, and is seen by another vulture, flying perhaps

MAN'S
ANIMAL
NEIGHBOURS

INTRODUCTION

Man's effects on the environment are usually destructive. Since prehistoric times he has cut down forests, ploughed plains and diverted or dammed streams, and in recent years has covered increasing areas with the concrete of towns, roads, factories and airports. Plants and animals have disappeared as man advanced and it is because of this that almost all countries have set up nature reserves, where relics of the formerly widespread fauna may survive.

Not all animal species, however, have been harmed by man. A few, more adaptable than most, have learned that nearness to man may mean food and shelter, and these creatures have accompanied him almost wherever he has travelled over the world. Sparrows, nesting in buildings and feeding on scraps and waste are a case in point. House Mice, which were native to Southern Europe are now found far beyond the warmth of their original home, for they have been carried by man to the far corners of the earth and survive in many of them only because of the shelter of human buildings. But these neighbours of man are rarely neutral in their relationship with him. Sparrows may become a pest in grainfields; mice are a much bigger pest in houses and stores, where they eat and damage food on a large scale. To an even greater extent rats, both the Brown or Norway Rat and the Black or Ship Rat damage food and stored products and other property and worse, spread diseases, of which the most dreaded is probably bubonic plague, the Black Death of the Middle Ages. Apart from these general cohabitors, man often offers to other animals bonanza conditions for their particular way of life. A granary, away from the normal predators, is a place where Grain Weevils can thrive; roof timbers, safe from the attacks of woodpeckers, can house wood boring beetles which may breed there and cause a great deal of damage. Woollen and other natural fibre cloths are a boon to those creatures whose normal way of life is to be scavengers of dead plant and animal material. Indespensable in their own environment, they quickly become pests in man's.

Those creatures which have discovered that close contact with man or his activities is advantageous to them have mostly become pests, but many animals live on the fringe of man's activities. These are by no means always harmful, and in many cases are to be welcomed for their beauty or the intrinsic interest of their presence. In parks and gardens, for instance, man has often created conditions approximating to those of old woodlands or of flowering meadows and many creatures have taken advantage of the fact. Birds, such as Titmice, or members of the Thrush family, loose much of their fear of man under these conditions and many species, including Flycatchers and even Owls, benefit from the setting up of nest boxes in which they may rear their young. Insects, such as butterflies and many colourful beetles may be found feeding on garden flowers and others, which are often overlooked, are present. A naturalist will often find great richness and variety of life in a garden or park, where animals have discovered the art of living alongside man.

WORMS

BRANDLING —
Eisenia foetida

In structure and life history the Brandling is similar to the other earthworms found in gardens and meadows throughout Europe (see page 183). It is a small worm, rarely reaching a length of more than 80 mm. and is most commonly found in compost or dung heaps. It may be recognised by its striped appearance, for as it extends to move, each segment of the dark red body may be seen to be banded with a yellowish colour. It feeds on plant and animal debris and is apparently distasteful to birds, for they usually avoid it. Nonetheless, it is very attractive to fishes, and is frequently used for bait by fishermen.

MOLLUSCS

GREAT GREY SLUG —
Limax maximus

The Great Grey Slug, found from Scandinavia southwards, is a harmless creature. Its food consists entirely of decaying plant and animal material and fungi and it never eats seedlings or damages living crops. As its name suggests, it is large, reaching a length of up to 200 mm. when extended. It is variable in colour, ranging from mottled yellow, through red-brown to almost black. One of the strangest habits of slugs is the use of their slime to make a rope from which they may hang if wishing to explore territory at another level. The Great Grey Slug also mates in mid-air, the two partners starting their courtship on a branch or wall and later dangling together from a thick twist of mucus.

ROMAN SNAIL —
Helix pomatia

The Roman Snail is found on the North Downs and other chalky places in southern England and across the Continent to Southern and Eastern Europe. It is one of the many animals which the Romans are credited with taking to Britain, but there is no certainty of this, although they were certainly cultivated by the Romans as a table delicacy. The species is still consumed in large quantities in France, and in Britain the populations of snails on accessible areas of the Downs dropped sharply after the 1939—45 war, when restuarateurs and others discovered that this continental delicacy occurred at home. The Roman Snail is unmistakeable among British species, for no other approaches it in size. Its almost circular shell may measure about 40 mm. across and the total length of the animal when active is about 80 mm. In general it lives in fairly warm, sheltered places where there is enough moisture to produce the moderately lush vegetation on which it feeds. As with all European land snails, this species is hermaphrodite; mating takes place in the summer months and the eggs, between 20 and 60 in number, are laid in a little pit in the soil. During the wintertime, Roman Snails hibernate, digging themselves into the top few centimetres of soft soil and closing the apperture of the shell with a cover of hardened mucus which they will also do in times of extreme heat or drought.

SPIDERS

GARDEN CROSS SPIDER —
Araneus diadematus

The beautiful orb web of the Garden Cross Spider is

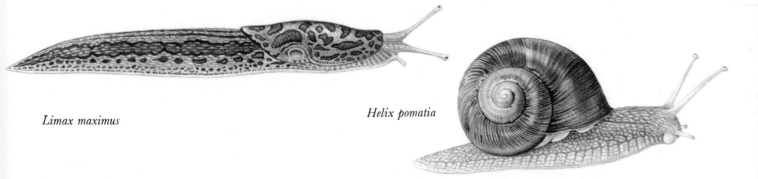

Limax maximus

Helix pomatia

a familiar sight throughout Britain and much of Europe in late Summer and Autumn. In spite of its name, this spider does not confine itself to gardens, but is common in a number of other habitats as well. The web is the home and food trap of the female spider, which measures up to 12 mm. long. She may be seen sitting, head downwards, at its centre, or hidden nearby at the end of a signal line. In either case she knows quickly if an insect has blundered into the web and she hurries down to bite it and swathe it in bands of silk before carrying it off, either to be eaten, or to wait in her food store. The male spider, which is smaller than the female, runs the risk at all times of being a meal rather than a mate, but does his best to prevent this by tweaking the web and swinging away on a thread of silk if its owner greets him too fiercely. No male stays for more than one mating, although he may mate later with other females. Several hundred eggs are laid, protected in a cocoon of yellowish silk which is placed in a safe nook where they pass the Winter. They hatch next Spring, but the little spiders are not adult until the following year.

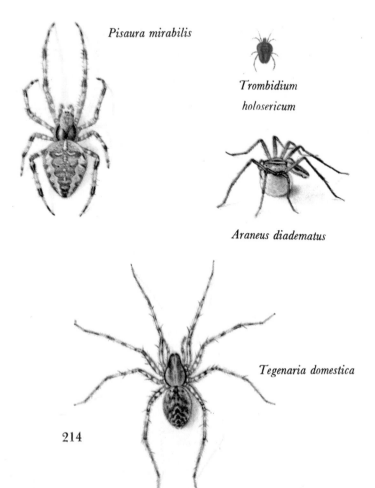

Pisaura mirabilis

Trombidium holosericum

Araneus diadematus

Tegenaria domestica

214

HOUSE SPIDER —
Tegenaria domestica

The dense white webs of the House Spider may be seen in neglected buildings throughout the world, for this is another species accidentally carried by man in his travels. It has a body length of about 10 mm., but the long legs make it seem larger than it really is. At one corner of the web is a tube, which is where the spider sits, hidden, waiting for flies and other small insects to become enmeshed, when she will run out and bite them. Her egg sac, which is a dirty white colour, dangles from a ledge, hung by a few threads of silk. Perhaps because this spider is usually found in the shelter of houses, it may be seen at all seasons of the year. It is long lived and probably survives for several years.

WOLF SPIDER —
Pisaura mirabilis

The female *Pisaura*, shown here with her egg sac, has no settled home or web, but hunts insects in the sheltered places where she is to be found. Her body length is about 15 mm., and she may often be seen on vegetation, sunning herself, with the two front pairs of legs close together and stuck out stiffly at an angle of about 30 degrees from the body. In early Summer the male goes courting. He first catches a fly, which he wraps in silk, but does not eat. When he crosses the track of a female he follows it and on finding her adopts a strange posture with the body tilted backwards as he holds the gift-wrapped fly out to her. While the female is examining and eating the offering, he mates with her, occasionally sampling a little of the present during pauses in the proceedings. The female carries the eggs with her for some time, but finally attaches the cocoon to a blade of grass, where she stands guard over it for a while longer.

HARVESTMAN —
Opilio paretinus

Often mistaken for a spider at first sight, this species is found over much of Europe in outhouses or damp and shady places as well as in open fields. The body is not clearly divided into two sections as with the spiders, and the eight legs are relatively longer and thinner. These can be shed by the Harvestman if there is any danger, and towards the end of the Summer most Harvestmen are minus at least one leg. When it is cast off the limb remains twitching for a while and this probably distracts a predator from the creature which is making its escape. Although without spider-like poison fangs, Harvestmen feed on other animals which they catch in the leaf litter or on the ground. There is no elaborate courtship before mating, as with spiders. The eggs are laid on the ground where they overwinter before hatching.

BOOK SCORPION —
Chelifer cancroides

The Book Scorpion is a tiny animal sometimes found in houses. It has huge, scorpion-like claws, but a rounded abdomen with no sting, like a true scorpion. Although only 3—4 mm. long, it is a hunter, seizing the young of such animals as Silverfish in its pincers which are provided with sensitive tactile hairs and poison glands. At mating time a male displays before a female, waving his pincers in the air. He deposits a packet of sperms on the ground in front of her, then retreats, beckoning her towards him. When she is standing above the spermatophore, he grips the base of her pincers in his and taps her with his first pair of legs, so that it becomes lodged in the correct place for fertilisation of her eggs to take place. The up to 40 eggs are retained by the female and the young are nourished by a milk-like fluid provided by her. Growth is rapid; moults take place in a specially constructed silk nest and the whole life cycle lasts one year.

VELVET MITE —
Trombidium holosericum

This little red creature, about 2,5 mm. long may sometimes be spotted on a warm, dry summer day. At such a time it may hunt for the tiny insects and their eggs which are its food, on dry paths or in open places near to houses. It is closely related to many other similar creatures, which are an abundant and important part of the life of soil and leaf litter.

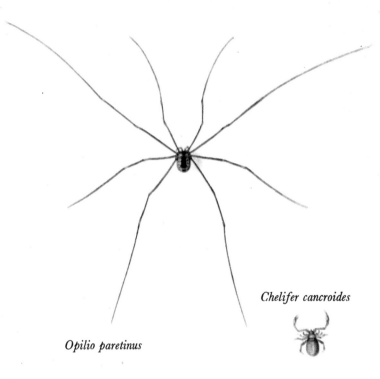

Chelifer cancroides

Opilio paretinus

CRUSTACEANS

WOOD LOUSE —
Porcellio scaber

Wood Lice are crustaceans and are rather less well adapted for life on dry land than are most of the other small invertebrates found round about houses. They have to remain in places where humidity is high, or they would be in danger of dessication, so they are, therefore, commonest in outhouses and gardens. This species can grow to a length of about 15 mm., but it cannot protect itself by rolling up, as can some of its relatives. It feeds on various sorts of decaying matter and can sometimes become a pest in gardens. —

INSECTS

SILVERFISH —
Lepisma saccharina

The shiny scales on the body of the Silverfish give it

Porcellio scaber

its name, for as it scuttles away from the light, its smooth, silvery appearance is fishlike. Although wingless, it is an insect, belonging to a primitive group descended from the earliest of insects, which had no power of flight. Some closely related species live out of doors; this one is found in houses, scavenging tiny fragments of starchy material, such as breadcrumbs or flour which may have been dropped on the floor of kitchens or bakehouses. Under normal conditions they do very little harm. If large numbers build up some damage may occur, the worst being when they attack the starch paste used in bookbinding.

COMMON COCKROACH —
Blatta orientalis

The illustration shows a female cockroach, carrying her eggcase or *ootheca*, projecting from her hind end. She measures about 27 mm. long and has much reduced wings. Her mate, which is usually slightly smaller, has rather larger wings, but is still unable to fly. There are pockets in Britain in which Cockroaches have not been recorded, but in general they are found throughout the world. It is probable that they originated in North Africa and spread to Britain in the sixteenth century. Certainly they are descended from tropical insects, for they cannot survive the Winter out of doors in cold climates. They are usually found indoors, in warm, dark places, from which they emerge at night to feed on an astonishing variety of objects, including any sort of human food, leather, paper, woollen clothing or hair. They damage and dirty more than they actually destroy and although they do not spread diseases as some flies may, they are nontheless undesirable pests. Each female produces several oothecae in her life. The young which hatch from them may take a year to reach maturity and survive for another six months after this.

GERMAN COCKROACH —
Blatella germanica

This creature, which has an almost worldwide distribution, certainly did not originate in Germany. North Africa is a likely source; it did not reach Britain until relatively recently and is still unrecorded in

a number of areas. It is a smaller insect than the previous species, averaging about 12 mm. in length. It has fully developed wings, but it normally prefers to run rather than to fly. It is to be found in buildings, especially in warm places such as bakeries, restaurants and kitchens, although during the summertime it may also be found on rubbish dumps. The ootheca is carried for several weeks before the young hatch; it is usually, but not invariably, dropped a day or so before this occurs, but the young are not, in any case, protected or helped by their mother. Under some circumstances unmated females can produce oothecae, but unlike the Common Cockroach in which this may occur, producing female young, in this case the eggs do not survive to the point of hatching. The nymphs take between 2 and 4 months to reach maturity, although in warm conditions it may take less than this. Before mating there is a complex courtship, involving antennal contact between partners. Before mating is accomplished the female feeds on a special secretion produced in hollows in the abdomen of the male.

MOLE CRICKET —
Gryllotalpa gryllotalpa

The Mole Cricket is found over most of Europe although it is rare now in Britain. It is a large, rather stout insect, averaging about 40 mm. in length, but it is rarely seen, for it spends much of its time underground, in tunnels dug with the huge shovel-like front legs. In dry soil, the burrows may be 30 centimetres or more underground, but in damp places they will be nearer the surface, for although the Mole Cricket can swim, it cannot survive in completely waterlogged conditions. It can fly, but is noisy and clumsy and normally only takes to the wing on very warm evenings. In Spring and Summer the male sits at the entrance to his burrow, singing his soft, monotonous, churring song and swaying from side to side should a female appear. After mating several hundreds of eggs are laid in an underground nest chamber. The female protects and licks them, which helps to prevent them from becoming mouldy. They hatch within a few weeks and complete their development during the next year, or in some cases the year after that. Mole Crickets feed underground, partly on roots and in some areas may cause substantial damage to crops such as carrots, but they also eat grubs and other invertebrates. They may themselves be food for a number of small flesh eaters such as Moles, from which they may sometimes escape by their habit of running fast, backwards, down their burrows.

CHICKEN LOUSE —
Menopon gallinae

Domestic animals carry parasites which may damage or weaken them. They may also get on to human hosts, where they can cause temporary discomfort. One such is the Chicken Louse, which measures about 1—1,5 mm. long and feeds on the blood and feathers of the birds which are its normal hosts. A heavy infestation may cause serious damage to young stock and anybody keeping chickens must attend to the cleanliness of their quarters and occasionally dust the birds with insecticide to combat this pest.

COMMON EARWIG —
Forficula auricularia

The illustration shows the male Common Earwig, which is distinguished by the strongly curved calipers at the end of his body; in the female these are almost straight. In both cases their main use seems to be in folding the large hindwings, which are used in the Earwig's rare flights, under the small protective flaps of the forewings. The Common Earwig, which is found widely in Britain and Europe, is largely nocturnal and during the daytime hides in crevices just big enough to accept its flat body. The adults hibernate through the cold part of the year and the female lays her eggs in the late winter months in a small chamber hollowed out in the soil. She broods and cares for them until the nymphs are independent, thus showing an early stage in the development of insect social life.

FIREBUG —
Pyrrhocoris apterus

The vividly coloured, 9 mm. long Firebug is widespread over most of Europe, except for the coldest areas.

Gryllotalpa gryllotalpa

Blatta orientalis

Blatella germanica

Forficula auricularia

Lepisma saccharina

Menopon gallinae

Pyrrhocoris apterus

Eurydema oleracea

In Britain, however, it has been recorded in various coastal localities, and is now known to survive in only one of these. It is a gregarious creature, usually found on lime trees or mallows in Europe and in Britain on tree mallow. It feeds on the host plant and sometimes on other insects. Firebugs overwinter in the adult stage and mate in late Spring, producing a single generation, which is mature by about August.

217

BRASSICA BUG —
Eurydema oleracea

This little bug, which measures only about 7 mm. long, is found rather rarely in southern Britain, where its food plants are usually wild members of the wallflower family, and it may sometimes also eat eggs and other small insects which it discovers there. In Europe it sometimes occurs on cabbages or related crops, where it may do considerable damage. A typical specimen is illustrated, but the colour is very variable, the males in particular often being darker.

BLACK BEAN APHID —
Aphis fabae and
PEA APHID —
Acyrthosiphon pisum

These animals are both small, soft skinned relatives of the bugs described above and like them have tube-like mouthparts with which they suck sap from the shoots of plants. Of the many European species of aphids the two figured are widespread pests of crops. They may be recognised as aphids by their small size, their long antennae and the two small spikes or cornicles at the hind end of the body. They are often found in very large numbers, some individuals winged and others, such as those illustrated, not. Through the summer months all aphids are females, which reproduce parthenogenetically. Each young female produced gives rise to daughters of her own within 8 to 10 days, although the exact time and the numbers born depend on the temperature. In the Autumn some males are born. Those of the Pea Aphid mate with the females on their host plant and eggs laid after this overwinter. In the Bean Aphid males and females fly to spindle trees, and the eggs are laid and overwinter there. Early in the Spring they hatch into female offspring which feed on spindle shoots, but later generations fly off and infest bean crops.

WOOLLY APHID —
Eriosoma lanigerum

Apple trees in springtime are often covered with what looks like dabs of cotton wool. These are the strands of waxy material covering another aphid, sometimes known as the American Blight. It has a life history similar to the previous species, except that overwintering eggs are laid on apple trees. A heavy infestation may do considerable damage, and the pest cannot easily be killed with insecticides, although it may be controlled by introduced parasites.

PARASITIC WASP —
Diaretus rapae

This tiny insect is a relative of the social wasps and bees, but its way of life is very different from theirs. In summertime large numbers of these animals are recognisable by their small size, their wasp-like appearance, with a narrowed waist, with long antennae and in the females conspicuous ovipositors. They may be seen on vegetation in gardens and in the countryside. Most are strongly host specific, as is the specimen illustrated, which injects each of her eggs into an aphid. When they hatch, the larvae feed on the aphids, killing them and pupating in the empty husks of their bodies.

BLACK GARDEN ANT —
Lasius niger

Evidence of the Garden Ant's activity may be seen in small heaps of fine grained earth in grass or by paths. The nest in which the colony lives is some way below the soil and this contains the queen, which is the mother of all the individuals in it, and the workers, which are her sexless daughters. Each Summer a generation of fully sexed males and females is produced. On a suitable day in late Summer, usually when the weather is warm but humid, males and females from nests over a large area emerge and embark on their marriage flight. After mating, the males soon die, but the females shed their wings and descend undeground to start a new dynasty. At first they live on the food reserves in their now useless wing muscles, but after the first workers have been reared, these take over, foraging for food of all sorts, attacking small insects and collecting plant material which they carry to the nest. They feed and care for further generations and for the queen, who from now on is nothing more than an egg-laying machine for the rest of her life, which may

Eriosoma lanigerum

Diaretus rapae

Lasius niger

Aphis fabae

Hister unicolor

Dermestes lardarius

Acyrthosiphon pisum

last for several years. One important food is honeydew, a sweet secretion produced by aphids and scale insects, which are in effect farmed by the ants and milked of this substance by them.

Hister unicolor

This shining black beetle, which measures 7—9 mm. long is common all over Britain and widespread in Europe. It is a scavenger and hunter of small creatures and may usually be found in compost heaps or cow dung, where these abound. They are also to be seen on the remains of dead birds or mammals. A number of closely related species are known; they are similar in shape, but mostly have splashes of red on the black elytra.

Trichodes apiarius

This beautiful beetle, which measures 9—15 mm. long has been recorded from Britain, in the past but there is little doubt that it no longer occurs there, although it is common enough in the warmer parts of the Continent. The adult may be found lurking around flowers in the early part of the Summer. It is not interested in the blossoms but in other insects which visit them, for it is a hunter and will attack any small creature. The larvae are generally found in the nests of solitary bees, where they feed on the grubs and pupae. They may sometimes enter the hives of Honey Bees, especially when these are in poor condition, and they attack not only the brood, but also weak adults.

Trichodes apiarius

Attagenus punctatus   Anobium pertinax

Aphodius frumentarius

BACON OR LARDER BEETLE —
Dermestes lardarius

The Larder Beetle is only 6—9 mm. long, but it has colonised man's world totally to its own advantage and although it probably originated in South East Asia, it now occurs almost wherever man lives. In the wild, the larva, which is covered with long black hairs, is a scavenger, attending to the last stages of destruction of the remains of other animals. As a neighbour of man it does particular damage to furs, hides and carpets and can wreak havoc among museum collections. In food stores it is especially damaging to bacon, smoked meats and cheeses. Under sheltered and warm conditions it can complete its life cycle in as little as six weeks, so high populations of these pests can build up very quickly. In houses, regular disturbance by cleaning will discourage them; in food stores etc., fumigation methods usually have to be used.

Attagenus punctatus

An example of how closely related insect species may differ in their ways of life is given here. *Attagenus punctatus* is found in the summer months in gardens and the countryside, feeding on nectar and pollen, which are essential to the females before they can lay fertile eggs. The pupae live in holes in wood, and the total life cycle is two or sometimes three years. *Attagenus pellio*, which is illustrated above, is similar in size and appearance, except that it carries only two pale spots on the back, is a major pest of hides, leather, and so on, and may be found in houses and warehouses where it completes its life cycle in as little as three weeks, if conditions are ideal.

Anobium pertinax

This little wood boring beetle, which is related to the common Furniture Beetle, may be found in fence posts and dead coniferous wood in gardens in Europe, although it does not live in Britain. It is rarely found in furniture or beams, for it prefers mouldy sap wood to the dry material used in houses. The life cycle lasts for two years. The female lays her eggs separately in old tunnels, bored by other insects in wood and the larvae feed and make new tunnels leading from them. When adult, the males announce their presence by tapping their heads against the tunnel walls, in a grouped sequence of 7—8 knocks at a time. Like the Death Watch Beetle in Britain, this habit has given *Anobium pertinax* a sinister reputation on the Continent, where in some areas it is known as the Dead Man's Clock.

OIL BEETLE —
Meloe proscarabeus

This fat, sluggish oil beetle, which measures 12—13 mm. long, may be seen over much of Europe crawling on the ground in early Spring. Apparently vulnerable, it is protected by an unpleasant irritating secretion and so goes unscathed. The female lays large numbers of tiny eggs in the soil — a necessity for few of them will survive the curious vicissitudes of their life history. The eggs hatch into very active, bristly larvae, which climb up the stalks of nearby vegetation. Here they wait for a particular kind of solitary bee and grasp its furry body when it visits the plant they are on. In many cases, no bee comes, in others they hang on to the wrong sort of insect and soon die, but those which have the good fortune to get the right sort of lift, are taken to the bee's nest, and there first raid a cell and eat an egg, then moult into another form, capable of feeding on the food intended for the bee's grub. It then goes into a resting stage, a feature known as hypermetamorphosis from which it emerges before entering true pupation. The life span of the Oil Beetle seems to be a long one, with two overwintering periods, one in the early larval stage and the other in the resting phase, before emergence as an adult.

CELLAR BEETLE —
Blaps mortisaga

Another species capable of protecting itself chemically, this time with an unpleasant smelling secretion is the 25 mm. long Cellar Beetle. This creature is found over much of North, East and Central Europe, but is rare in Britain occurring only in Scotland, although a re-

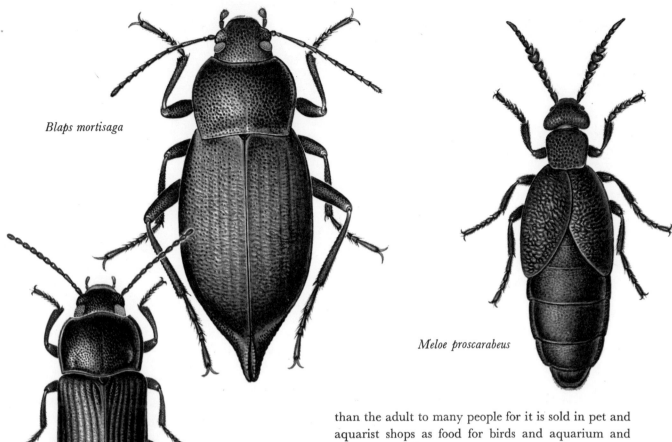

Blaps mortisaga

Tenebrio molitor

Meloe proscarabeus

lated form is somewhat more common in the south of the country. In spite of its rather unprepossesing appearance it is quite harmless. Strictly nocturnal, it spends the daylight hours in hiding, preferably in damp, cool places, like cellars, but at night emerges to feed on any food scraps which it can find. Its association with neglected, often dirty houses has led to superstitious beliefs in parts of Europe, where it is credited with being a herald of death; a reputation which is quite unjustified.

MEALWORM —
Tenebrio molitor

The larva of the Mealworm Beetle is better known than the adult to many people for it is sold in pet and aquarist shops as food for birds and aquarium and terrerium animals. If the hard, yellow coloured grubs are kept, they pupate and finally emerge as beetles 14 to 17 mm. long. They are rarely seen in ordinary houses, but are a commonplace pest all over the world in warehouses storing food, mills and places where domestic animals are kept. They need warmth and dryness for survival and over most of Europe are unable to live outside, but the larvae are almost completely ommniverous and will survive on the most unpromising diets, such as chicken droppings, as well as the more obvious foods such as bran, offal and waste from slaughterhouses. The damage done by Mealworms is very great, but it can be minimised by reducing access to places where waste materials are kept in which they might breed. They may also be combatted in food stores etc. with modern insecticides.

Aphodius frumentarius

On sunny spring days, often as early as the beginning of March, large numbers of these 5—8 mm. long dung beetles may be seen, enticed from their winter hiding

221

Phyllopertha horticola

Amphimallon solstitialis

places by the warmth of the spring sunshine. They are common in Britain and are found all over Europe, including Scandinavia and occur in Asia and North America as well. The females lay their eggs in cow or horse dung which is the food for the larvae.

SUMMER CHAFER —
Amphimallon solstitialis

This beetle 14—18 mm. long is a small relative of the Cockchafer and like it may be a pest, although in Britain it is only occasionally very abundant and is in any case rarer towards the north. In Europe it is widespread from central Scandinavia into Asia Minor. During the daytime the adult beetle rests, but at night emerges to feed on leaves of trees, sometimes damaging fruit trees in gardens. In July the females lay their eggs in soft soil. The larvae remain underground feeding on the roots of grasses or other plants, including cereal crops or vegetables.

COCKCHAFER —
Melolontha melolontha

An alternative name for this beetle is the May Bug, for it is active in the early part of the Summer. Although nocturnal it is often seen, for it is attracted to light and may come blundering in to uncurtained rooms at night, its loud whirring flight proclaiming its identity. It is common in south and midland Britain, but less so in the north. In Europe it is found from south Sweden south and eastwards into Asia Minor. In Britain it rarely reaches pest numbers; in Europe it is a major problem, for the adults defoliate trees often causing great damage. In June the females lay several batches of eggs, totalling about one hundred in all, in soft ground in meadows or gardens. The grubs feed at first on decaying substances in the earth, but having overwintered deep in the soil, return to the surface and feed on the roots of grasses, field crops and young trees. During this part of their life they are known by the descriptive name of White Grubs for their heavy curved white bodies are nearly helpless to do anything other than burrow and eat. They have

enemies among the insectivores such as moles and shrews and birds (such as Starlings) which probe the ground with their long beaks. Frequently however, enough survive to do a great deal of harm to crops, sometimes even killing young trees. The complete development of the larvae takes 2—3 years in the south and may be longer in the north.

JUNE BUG —
Phyllopertha horticola

This beetle which is a small relative of the previous species measuring 8—11 mm. long is common and widespread in Britain where it occasionally reaches pest proportions. It is unusual in that it is known by a number of vernacular names, including 'Field Chafer', 'Bracken Clock' and 'Fern Web'. However it rarely occurs in the numbers recorded for mainland Europe, where population densities may reach a peak of up to 130 larvae to a square metre of ground. The adults feed on a wide variety of plants, including the leaves of roses, soft fruit and fruit trees, which they may damage severely. The larvae, which hatch from eggs laid in the ground in early Summer, feed largely on the roots of trees and shrubs. They overwinter deep in the soil and continue feeding next Spring, pupating in May.

ROSE CHAFER
Cetonia aurata

Rose Chafers are far less common, than the previous species. They may occasionally be seen in some numbers in southern England, but become rare towards the north. In Europe they are widespread south and eastwards from central Scandinavia to as far as Asia Minor and Siberia. The adult, which measures between 14—20 mm. long, is more active in the daytime than its relatives and may, as its name suggests, be found most commonly on rose trees, either wild or cultivated, although it may infest other shrubs as well. The eggs are laid in decaying tree stumps where the larva takes one year to complete its development. It pupates in the soil in a pupal chamber made of wood fragments and soil, cemented together with a sticky secretion.

HOUSE LONGHORN —
Hylotrupes bajalus

This beetle 10—20 mm. long, which in the wild helps to break down the wood of dead coniferous trees, has become a major pest of softwoods used in building construction. The female lays her eggs in rafters or beams and the larvae burrow at first superficially and then deeper into the wood, destroying it almost completely. The wood needs to be damp, but if the humidity is too high the eggs probably become mouldy and do not hatch. It is widespread in Europe from Scandinavia to North Africa, but in Britain is known only from a few localities in southern England. It has been introduced to North and South America, Africa and Australia. Temperature affects the length of time the grub takes to develop. In South Africa one year is adequate; in Europe three years is a minimum; eleven has been recorded as a maximum. In Britain six years is the average time.

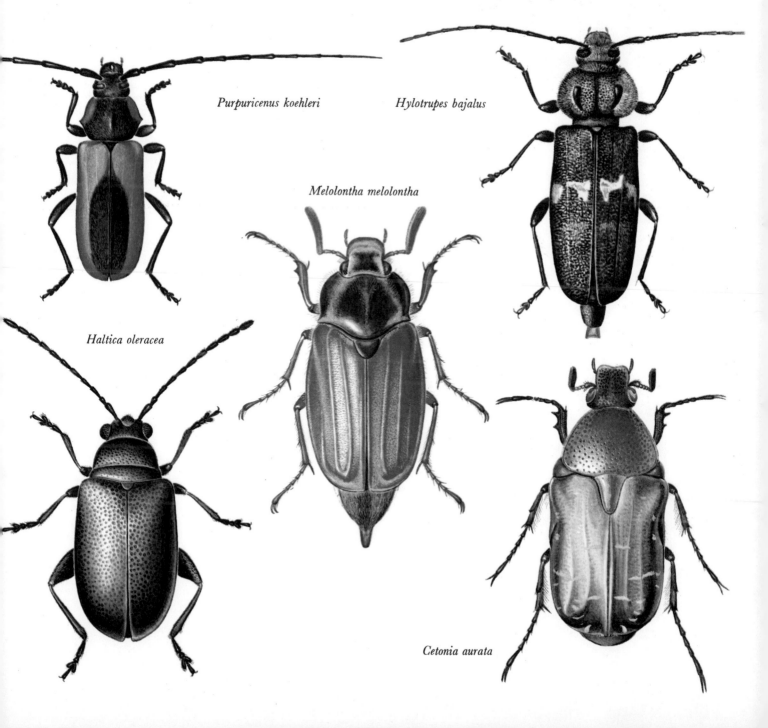

Purpuricenus koehleri

Hylotrupes bajalus

Melolontha melolontha

Haltica oleracea

Cetonia aurata

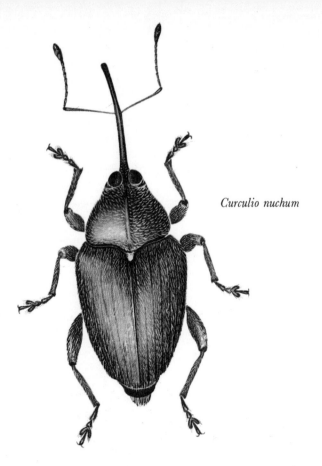

Curculio nuchum

Purpuricenus koehleri

This brightly coloured beetle, which measures up to 20 mm. long, is not native to Britain or Northern Europe and is rare in Central Europe. The larvae develop in peach, plum or pear trees, and the adult is on the wing during the summer months.

FLEA BEETLE —
Haltica oleracea

This Flea Beetle is about 3—4 mm. long and may be found over most of Britain and Europe during the summer months, feeding on various plants especially those of the cabbage family. It lives in colonies and is often very destructive to both leaves and flowers. When disturbed it leaps away in all directions, a habit from which its vernacular name is derived. The larvae which are mainly greenish in colour, are covered with curious spine-like projections.

NUT WEEVIL —
Curculio nuchum

This strange looking small beetle which is widespread

in Britain and Europe, is associated with hazel-nut trees and may occur in gardens or in woods. The snout carries a pair of hard jaws at its tip, and is longest in the female who uses it to bore a hole through the shell of a young hazelnut. She then lays an egg in the nut and the larva feeds on the tissues of the developing kernel. The original hole closes up and the grub is safe from all harm. Eventually it falls to the ground and the larva gnaws a small hole through which it escapes to the outside world. It pupates in the soil, emerging as an adult during the next Summer. Although the larva has few enemies, the adult may fall prey to a solitary wasp, which stocks her nest with them for her grubs to feed on. The wasp immobilises the heavily armoured beetles by stinging them in the soft places between their joints.

LACE WING —
Chrysopa vulgaris

The elegant Lace Wing, which measures about 20 mm. in length is often found in houses during the winter months. It may be known by its large delicate wings and green body and its big golden coloured eyes. In hibernation the green colours fade and are replaced by a brown shade which vanishes in the springtime when the insect leaves its shelter for gardens and fields. Here it hunts aphids, mates and reproduces. The eggs of Lace Wings are usually protected under a leaf and each egg is laid at the end of a long stiff stalk, which is probably a protection against predators which would destroy it if it were laid directly on the leaf blade. The larva is an active grub with huge mandibles, and like the adults, it feeds on aphids.

GOAT MOTH —
Cossus cossus

The Goat Moth is so called because the caterpillar is reputed to smell like a billy goat. This however is difficult to check since the creature burrows deeply as it feeds on the wood of willows, elms, ash and birch trees. When fully grown, after three or four years development, it may wander away from the tree in search of a suitable spot for pupation. On emergence, the adult moth is well camouflaged and unlikely to be noticed, although the female has a wingspan of about 90 mm. The eggs are laid in cracks in the bark of old but not dead trees, particularly isolated specimens in parks or gardens. The caterpillar soon hatches. The destruction caused by large numbers of these moths can be very great, for the caterpillar reaches a length of 90 mm. and gnaws holes which are oval in shape and up to 15 mm. wide, penetrating into the heart of the tree. Goat Moths are found in suitable places throughout Britain, except for the most northerly areas. On the Continent they are found almost throughout Europe and into Asia and North Africa.

LEOPARD MOTH —
Zeuzera pyrina

The Leopard Moth occurs mainly in southeast England, and in Europe and Asia is widespread as far east as Japan. It has been introduced into America and has established itself there, although it does not seem to have become a pest yet. The female lays her eggs in crevices in many sorts of hardwood trees, where the caterpillar tunnels for the next two or three years. It returns to near the surface of the bark to pupate, emerging as an adult in about midsummer. The female has a wingspan of about 65 mm., the male about 40. They are very much attracted to bright lights and are quite often caught in towns, including London.

BAG WORM —
Pachytelia unicolor

Visitors to parts of Southern Europe may sometimes notice curious cocoon-like structures, decorated with bits of leaves and grass seeds in the long wayside grasses. This is the protective case of the Bag Worm, a rarely seen moth, for the male which is small and dark in colour is night-flying and has a very short adult life. The female never leaves the sac which she spun first while still a caterpillar, for her wings and sense organs are degenerate. Here she mates and lays her eggs and the caterpillars once they hatch must force their way from the bag before they can start their independent life. In some parts of the tropics Bag Worms of related species are agricultural pests.

HORNET CLEARWING —
Sesia apiformis

Many insects protect themselves from predators by bearing a resemblance to creatures well able to look after themselves, by their sting or nasty taste. Hover Flies and many kinds of beetle for example, are superficially so like wasps as to deceive attackers, even including man, who will normally leave them alone. The Hornet Clearwing, with a wingspan of 40 mm., is the largest British representative of a group of moths which look at first sight like wasps or their relatives. This insect which is found in Europe and mainly in eastern and southern Britain, feeds as a larva in the roots and lower part of the trunk of poplar trees, taking two or probably three years to become fully grown. It pupates in a cocoon made of wood scrapings and silt which is placed

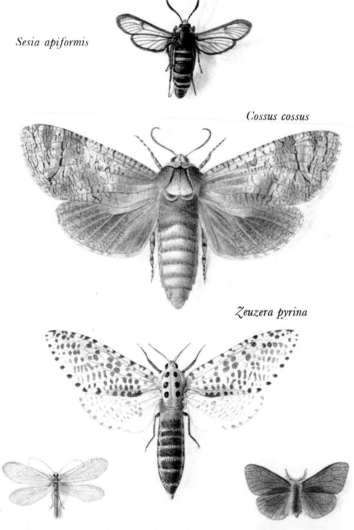

Sesia apiformis

Cossus cossus

Zeuzera pyrina

Chrysopa vulgaris

Pachytelia unicolor

on the bark of the tree or in the soil and emerges usually early on a sunny day in May or June. Because it has few scales on its wings, the adult moth looks very much like a Hornet, and its wing beat matches that of the large wasp and makes a buzzing noise which is a further warning to would-be predators.

GREAT EMPEROR MOTH —
Saturnia pyri

This handsome insect which has a wingspan of up to 150 mm. is Europe's largest moth. It does not occur in Britain or any of the cooler parts of Europe, but is restricted to the south and warmer areas. The adults are on the wing in May, flying chiefly on the edge of forests, in orchards, parks or large gardens. Here the females search for apple, pear or sloe trees, or failing these, willows on which to lay their eggs. The caterpillar which measures over 100 mm. long when fully grown is coloured green with pale blue or pink warts and some red markings on its sides and back. It carries a large number of stiff bristles which it can rattle when disturbed or picked up. This must astonish most predators, which may drop the insect in suprise and so let it escape.

LAPPET —
Gastropacha quercifolia

The female Lappet Moth may have a wingspan of over 70 mm., though her mate is rather smaller. It occurs mainly over Europe and Northern Asia and in Britain is found mainly in the south and east where it may be seen during the summer months. The females lay large numbers of eggs in small batches on the leaves of several species of trees, including orchard trees as well as hawthorn and willow. In Britain they are never sufficiently numerous to become a pest; on the Continent they may become so. The insect gets its English name from the caterpillar which has fleshy warts or lappets down its sides.

PALE TUSSOCK —
Dasychira pudibunda

The pale Tussock Moth, which may in the female have a wingspan of over 50 mm. is widespread in Europe and Northern Asia, as far as China and Japan. In Britain it is most common in the south, and although it has been recorded from the northern counties it is very rare there. It is on the wing in May and June and on the Continent a second generation may appear in the Autumn, although this is not so common in Britain. The moth, which used to be known really as the 'Hop Dog' because it was often found in hop fields, flies at night and rests, often in bracken or in nearby woodland, by day. The caterpillar feeds on hops, though nowadays this crop is usually protected by insecticide; it is also found on many trees including oak, hazel and birch. It is a striking looking creature, brightly coloured and hairy, with shaving brush like tufts of yellow or reddish hairs along its back.

WHITE ERMINE —
Spilosoma lubricipeda

The White Ermine Moth is to be found from Britain eastwards across Europe and into Asia. It is variable in coloration, some forms being more heavily spotted and one, found in northern Britain, having orange coloured forewings. The moth is on the wing in the early Summer and is frequently attracted to the lights of houses. The caterpillar which is covered with long, dark hairs, with a reddish line down the middle of the back is to be found on many sorts of low growing herbaceous plants in August and September and seems to show little preference among them. The chrysalis, which is protected in a cocoon of silk and hairs, is often attached to fence posts at ground level.

MAGPIE MOTH —
Abraxas grossulariata

The Magpie Moth, which has a wingspan of about 40 mm. is highly variable in the relative amounts of black and yellow colours on the forewing, though the typical form is illustrated here. It occurs through Britain, Europe and temperate Asia, and may be a major pest of gooseberries and currants, which are a favourite food of the whitish-coloured caterpillar, alth-

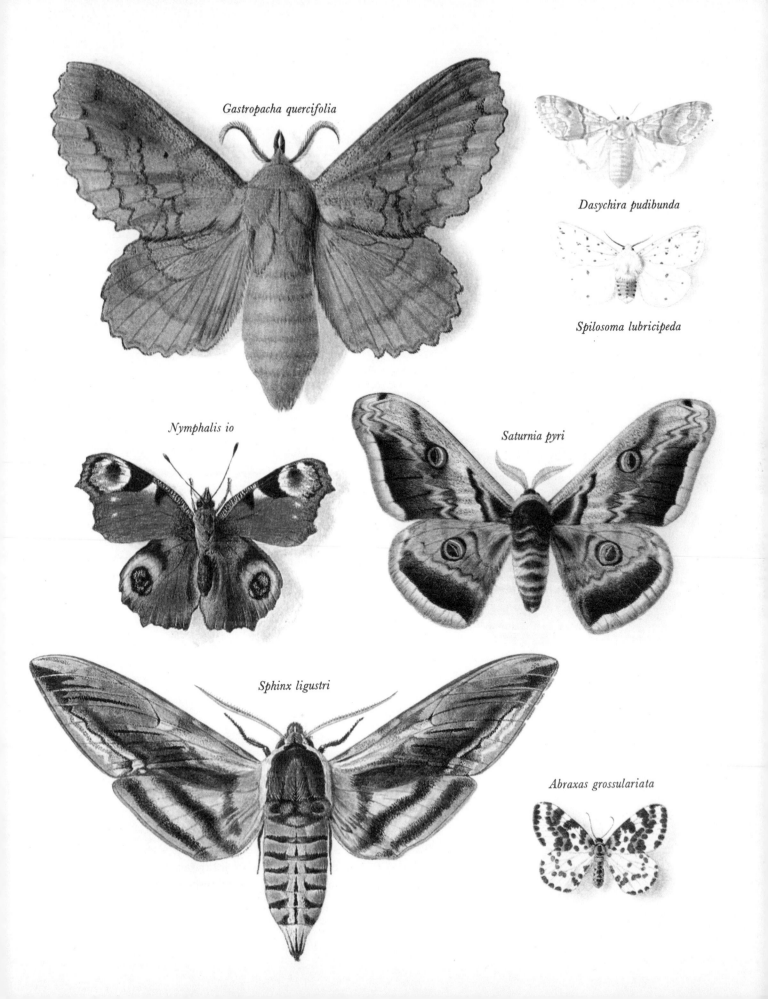

Gastropacha quercifolia

Dasychira pudibunda

Spilosoma lubricipeda

Nymphalis io

Saturnia pyri

Sphinx ligustri

Abraxas grossulariata

ough many other plants, including hedgerow species and even heather are recorded as being eaten by it. The moth is normally on the wing from middle to late Summer; the caterpillar begins its development in the year it was laid as an egg but usually hibernates through the Winter, pupating early in the next year. This species apparently tastes unpleasant and is normally rejected by birds and other predators.

PRIVET HAWK MOTH —
Sphinx ligustri

This is one of the commoner Hawk Moths to be found in Western Europe, although it is rarer in the south and east of the Continent. It is the largest of the British native species with a wingspan of over 110 mm. It may often be seen in the midsummer months visiting long-tubed flowers, such as honeysuckle for their nectar, even when these are in suburban gardens, for here privet hedges abound on which the eggs may be laid. But in the countryside a variety of other hedgerow plants may be eaten by the larvae.

PEACOCK —
Nymphalis io

The Peacock is one of the commonest and most easily recognised of British butterflies, and is also found throughout Europe and most of temperate Asia. Any park or garden containing buddleia bushes will attract them during the late Summer and it is likely that these same insects will then hibernate in a nearby shed or even a little-used room in a house. They will emerge from here in the first warm spell of the spring-time and at this time may be seen in their elaborate courtship-chasing flights. The eggs are laid on the underside of stinging nettle leaves, and the caterpillars stay together under the protection of a silk tent which they spin anew as they move to other plants. They pupate away from the nettle bed.

SMALL TORTOISESHELL —
Aglais urticae

The Small Tortoiseshell is to be found commonly in Britain and Europe and through Asia to Japan. It also hibernates through the winter months in the shelter of houses and outbuildings from where it emerges in the early springtime, to lay its eggs on the leaves of stinging nettles. The yellow and black caterpillars spin a protective tent on which they may lie and sunbathe when not feeding. Their growth is rapid and by mid-summer they have emerged as adults. These then produce another generation which survive the Winter. Many colour variants exist on the basic pattern shown in the illustration; the undersides are heavily camouflaged so that the bright butterfly seen in flight by a predator disappears as it drops to rest.

COMMA —
Polygonia c-album

This insect gets its name from a white comma-like mark on the underside of the wings, which are otherwise camouflaged with dull brown colours. The ragged edge of the hindwings is a further camouflage device, for with its wings closed the butterfly could be mistaken for a torn leaf, which has survived the Winter's storms. Like the two previous species it is common in parks and gardens during the summer months, and adults of the second generation often hibernate in houses. The caterpillars feed on several plants including hops, nettles and elm but the British population may be swollen each year by migrants from Europe and even from North Africa, for the species is widespread through the non-tropical parts of the Old World.

RED ADMIRAL —
Vanessa atalanta

This beautiful butterfly, with a wingspan in the female of 70 mm. is to be seen throughout the summer months in Britain and much of Europe, but north of the Alps it is unable to survive the Winter and the stocks are replenished each year by migration from warmer areas. A series of generations of butterflies is produced through the Summer, the caterpillars living singly in silk tents on their food plant, which is stinging nettle. At the end of the warm season the last generation may attempt to migrate southwards towards the Mediterranean. The adults, in spite of their intolerance of cold weather, often fly well into the evening and may like moths, be attracted to light.

SCARCE SWALLOWTAIL —
Iphiclides podalirius

This handsome member of the Swallowtail family occurs in Britain only as a rare vagrant, although it is found through most of Europe, North Africa and temperate Asia. Two broods are produced in the course of the summer months, the caterpillars feeding on the leaves of sloe, or more frequently cultivated fruit trees in orchards or large gardens. It lives mainly in lowland areas, is double-brooded, insects of the later brood having paler ground colour to the wings but more dark on the hind wing than the first brood. This brood in any case usually has a less intense colour than the illustration shown here.

LARGE WHITE —
Pieris brassicae

The Large White Butterfly is found throughout Europe and most of temperate Asia and probably is the best known butterfly in Britain for it occurs almost as commonly in the towns as in the countryside. A female has even been observed laying her eggs on cabbages in a London greengrocer's shop. The specimen illustrated is a female of the spring brood. The males have no black spots on the wings and in the summer brood both sexes have the wing tips more completely black. The females have a wingspan of up to 65 mm. but as in most butterflies the males are slightly smaller. The adults feed only on nectar, but the food plants of the caterpillar comprises many species of the wallflower family, both wild and cultivated. The latter includes all the cabbage-like

Aglais urticae

Polygonia c-album

Vanessa atalanta

Iphiclides podalirius

Pieris brassicae

Aporia crataegi

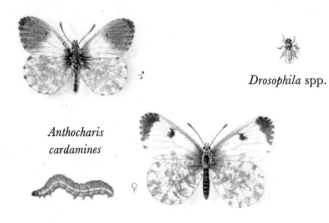

Drosophila spp.

Anthocharis cardamines

plants and this butterfly is an important pest of these crops. The caterpillars which are greenish yellow are social at first and at all stages seem to be distasteful to birds which rarely eat them. They are, however, kept in check by a parasitic wasp, which lays its eggs in the larva where they develop at its expense.

BLACK VEINED WHITE —
Aporia crataegi

This handsome butterfly with a wingspan of over 60 mm. in the female, is to be found on the wing in early Summer through much of Europe and Western Asia, but the numbers seem to vary greatly from one year to another. In Britain, where it was once common in the southern counties it is probably now extinct, although there is some hope that it may have survived in Kent. The food of the caterpillars is the leaves of hawthorn, sloe or various fruit trees and in some areas this species may be an orchard pest. The caterpillars overwinter gregariously in an early stage of development under a thick protective silk web spun in a tree. At first they feed together, but later wander off alone to complete their growth before pupating.

ORANGE TIP —
Anthocharis cardamines

Only the male of this species fits its name, for the female shown in the lower illustration, has black tipped wings. It is found throughout Europe except for the far north and the south of Spain. In Britain it is one of the early summer species, for it overwinters in the chrysalis and emerges in about late April or May At this time it may be seen fluttering along the hedgerows, looking for plants of Lady's Smock or Jack-by-the-Hedge on

which the eggs are deposited singly and where the caterpillar feeds, not on the leaves, but the seed pods.

HOVER FLY —
Syrphus ribesii

Many species of hover fly are common throughout Britain and Europe during the summer months. They may be recognised in general by their habit of hovering, wings moving so fast as to be practically invisible, in one spot, darting to another and suddenly reappearing at the first place. Many, such as the species illustrated here, are wasp-mimics and although they cannot sting, or protect themselves in any way, they are avoided by many predators including man. This fly and its close relatives are valuable to gardeners and horticulturalists because the larvae, which are sightless, legless grubs, have an insatiable appetite for aphids and a single *Syrphus*, may in the ten days of its larval life dispose of several hundred of them. After pupation which lasts another ten days the adult emerges. This is not a carnivore but feeds on pollen and nectar and is of importance as a pollinator.

CHERRY FLY —
Rhagoletis cerasi

The small, (3—4 mm. long) brightly coloured Cherry Flies are on the wing from mid-May to the end of June through much of Central and Southern Europe. Although they may occasionally reach Britain they do not breed there and it is probable that the climate is too damp for them to overwinter. Nonetheless all cherry imports are inspected for their presence, for this insect is a major pest of orchards on the Continent. The females lay their eggs in unripe fruit, which the larvae damage as they feed. Their activities cause premature colouring of the cherries; this attracts birds which act as a partial control of the pest, but may themselves do damage to the fruit. The larvae drop to the ground to pupate and overwinter there.

FRUIT or VINEGAR FLY —
Drosophila spp.

Several very similar species of *Drosophila* may be found

throughout the summer months in places where fruit or other materials may be fermenting. In general they may be recognised by their small size — about 3 mm. long — their stout yellowish bodies and their bright red eyes. Their flight is slow and rather laborious, with the abdomens hanging down somewhat. Their life cycle is extremely short and largely because of this and the fact that some of their chromosomes are extremely large they have been used by geneticists in the study of the inheritance of physical characteristics.

HOUSE FLY —
Musca domestica

The House Fly is content to stay inside houses and may sometimes be seen, flying round lampshades and other articles which form a focus for its activities. House Flies have been taken by man to every corner of the earth. A female House Fly will lay her eggs in almost any kind of rotting refuse and under ideal conditions the larvae may complete their growth in two days, although in colder places this may take up to eight weeks. After a short pupation the adult fly, which is about 8 mm. long, with a wingspan of 15 mm. emerges. It feeds on liquids which it sucks up; solids such as sugar, it first dissolves with a drop of saliva. Because of their habits House Flies may transmit bacteria and disease organisms from their egg laying places to food, should this be left uncovered, but in many places this danger has been reduced by the use of modern insecticides. These have made it a comparatively rare species in houses in Britain.

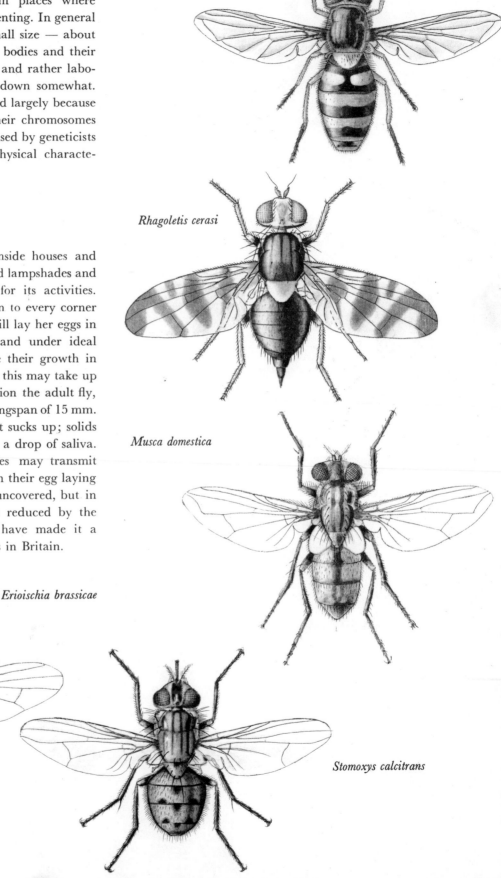

Syrphus ribesii

Rhagoletis cerasi

Musca domestica

Erioischia brassicae

Stomoxys calcitrans

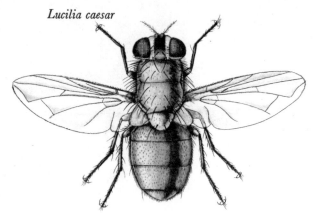

Lucilia caesar

Sarcophaga carnaria

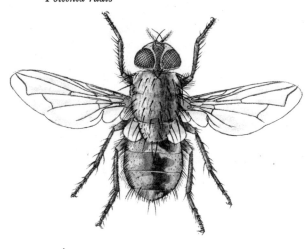

Pollenia rudis

Pulex irritans

STABLE FLY —
Stomoxys calcitrans

This fly is similar in size and general appearance to the House Fly, but may be distinguished at once by the forwardly projecting mouth parts, which form a strong piercing and sucking proboscis. Both sexes feed on blood, usually that of domestic animals, but sometimes they attack humans. They sometimes enter houses, but are usually found where horses or cattle are kept and the larvae develop in straw or debris which contains the animals' droppings.

CABBAGE ROOT FLY —
Erioischia brassicae

A number of species of small flies seem in many cases to have abandoned the wild plants which were their original food and taken to feeding largely on cultivated crops. One such is the Cabbage Root Fly, which although only 6—7 mm. long, is the source of much damage in cabbage, turnip and raddish fields. The harm is done by the larvae which tunnel into the roots and weaken the plant or in the case of turnips etc. ruin the crop.

GREEN BOTTLE —
Lucilia caesar

Green Bottles are somewhat variable in size and in colour which changes in intensity with age. They rarely enter homes, but prefer to find a dead animal where the eggs are laid, usually in a shaded part. The larvae produce a liquifying material, which allows them to suck up their partly digested food and when they are fully grown pupate in the soil. Like House Flies they have followed man all over the world, and in Australia particularly have become a serious pest, attacking wounds on sheep and cattle.

FLESH FLY —
Sarcophaga carnaria

This fly which is about 14 mm. long may be seen in places where any sort of carrion is to be found, for this is the food for the larvae. These are produced alive by the female, which does not lay eggs and like the Green Bottle grubs they also liquidise their food. They pupate in the earth, at a depth of several centimetres and escape when they emerge by means of an inflatable sac on the head.

CLUSTER FLY —
Pollenia rudis

In the wintertime Cluster Flies may be found hibernating in huge numbers in lofts or attics. They measure about 10 mm. long. As soon as the weather becomes warmer they leave their winter quarters for the outside world, where they are important as pollinators of some of the flowers of the early Spring. The larvae are parasitic on Earthworms.

HUMAN FLEA —
Pulex irritans

In general Human Fleas are found throughout the world, associated with man who is responsible for their spread. It is not easy to identify fleas and since cats, dogs and birds all have their own species which may be found on man at least occasionally, they will be dealt with jointly here. All fleas are small (2—4 mm. long) wingless and flattened from side to side so that they can creep easily through the dense fur or feathers of their hosts. All are parasitic on warm-blooded animals and all have hind legs which enable them to leap out of trouble. The adults feed on blood, which they suck from their hosts, and in doing so may spread disease organisms, the most dreaded of which is the bubonic plague, although this has not occurred in Europe for many years. The eggs are not laid on the host, but are dropped in bedding, or carpets etc. and here the flea larvae, which are legless but active grubs, feed on detritus of all sorts.

AMPHIBIANS

Bufo bufo

Corvus monedula

COMMON TOAD —
Bufo bufo

The Common Toad is widespread over Europe and Asia but is less abundant in Britain than formerly, because of changes of habitat and the use of agricultural chemicals which often poison the water in which it breeds. The adults live on land. Slow and clumsy, they would soon be wiped out by predators but for the poison glands in their skin, which protect them. They flock to breeding pools in the springtime, where the females lay up to 6,000 eggs in a double string which is wound round the water plants. The tadpoles metamorphose within the season, but the little toads are not sexually mature for several years. In Britain it is unusual for a toad to grow much above 80 mm. in length; in Southern Europe specimens of twice this size may occur. With the exception of certain distasteful insects, toads feed on any living thing which they can swallow including young mice, snails and ants. The enemies of toads include many birds, the Grass Snake and a parasitic fly, in which the larvae invade the nasal cavity and from there eat the skull tissues. In general, however, these nocturnal, hibernating creatures survive for many years.

BIRDS

JACKDAW —
Corvus monedula

Jackdaws are resident through much of Europe and

233

North Africa, and they spread into the cooler parts of the Continent in the summer months, so the populations are partly migratory. Large (about 330 mm. long), strongly social in behaviour and, as birds go, intelligent, the Jackdaw has discovered that living near to man may be advantageous, and is to be found many towns and villages, as well as in open farmland. They will eat a wide range of plant and animal food and are seldom at a loss for nourishment. Their natural nesting places are on cliff faces and in holes in trees. Old buildings make excellent substitutes for these and Jackdaws are often to be seen around dilapidated buildings where the female will incubate 4—6 bluish-green eggs in a secure if untidy nest. The young remain with their parents after they are fledged. Outside the breeding season Jackdaws often form mixed flocks with Rooks and Starlings.

STARLING —
Sturnus vulgaris

Starlings measuring about 215 mm. long are resident birds throughout Britain and most of the more temperate parts of the Continent and Asia Minor. Large numbers migrate into Scandinavia and North East Europe in the summer months, and Spain and the hotter parts see them in the wintertime. They are true neighbours of man, and have been taken to America and Australia where they are often regarded as a pest. Many Starlings feed in the country, or the suburbs, probing lawns with their long beaks to get at the juicy grubs which are their chief food. Outside the breeding season they form flocks which congregate at roosting places, many of which are in the middle of towns. A big Starling roost may contain more than a million birds, and as they come in to rest the great flocks wheel and circle in a silence unbroken except for the sound of their wings. As soon as they have settled hubbub breaks out, each bird chattering and squealing and in London this din can be heard even above the noise of traffic in Trafalgar Square, which is one of their major roosts. Starlings may cause damage to buildings when they roost in such vast numbers; in the countryside they often break the branches of trees and the thickness of their droppings sterilises the ground where they fall. Starlings are

great mimics and their calls and song often include phrases taken from other birds. In Britain a Starling may occasionally be heard giving the unmistakeable musical whistle of the Golden Oriole; proof that this is a continental bird which has migrated.

CRESTED LARK —
Galerida cristata

Although Crested Larks are rare visitors to Britain, they are found through much of Europe south of Scandinavia often living in small towns and villages. About 165 mm. long, they may be recognised by their plump form and crests as they trot along, often in small groups, scavaging close to houses or rubbish dumps. In flight the sandy colour of the underside of the wing is a good identification characteristic. Like many town animals they have become very tame.

HOUSE SPARROW —
Passer domesticus

In the days when horses were the main means of transport, House Sparrows were abundant in towns, feeding on grain spilt from nosebags. Nowadays, although still common in the suburbs and seldom seen far from human habitation, in the country the decline of the horse has meant reduction in Sparrow numbers. It is a smart, cheeky bird, 145 mm. long and always ready to exploit any possibility offered by humans in the way of food or nest sites. It nests in holes, making an untidy jumble of bits of dry grass, string, feathers and any other more or less soft material. In spite of this, recent research has shown that the Sparrow is related to the Weavers which make some of the most elaborate and beautiful nests of all birds. The female, which lacks the black bib of her mate, lays her first clutch of eggs in April and may produce four or more broods in a season. Sparrows are found throughout Europe, although their dependence on man is such that there are few in really remote areas. They have been taken by man to all parts of the world and are now inhabitants of America and Australia, where they are less than welcome immigrants, for they can damage grain crops and in some places have ousted the native birds from the area.

Fringilla coelebs

Galerida cristata

Sturnus vulgaris

ORTOLAN BUNTING —
Emberiza hortulana

The Ortolan Bunting is a summer visitor to much of Europe other than the western fringe and occurs in Britain only as a passage migrant. It is associated with cultivated areas and gardens, but tends to be shy and secretive, so may often be overlooked. The nest is made on the ground, and the 4—6 young are fed by both parents on small insects and other invertebrates, although the adults eat principally seeds.

Passer domesticus

Fringilla montifringilla

Emberiza hortulana

CHAFFINCH —
Fringilla coelebs

The brightly coloured cock Chaffinch and his more sombre mate are one of the most attractive of European birds and among the commonest, for they nest in all but the most northerly parts of the continent, from sea level up to 1,500 metres. In the Winter Chaffinches from most of Scandinavia and the north eastern parts migrate southwards, but return as soon as the ground begins to unfreeze. The Chaffinch's nest is a neat round construction of moss and lichens lined with feathers and hair placed in the fork of a small tree or shrub. Here, in about April, the female lays 4—6 brownish white eggs, freckled and streaked with rusty brown. They are incubated for about 14 days, after which both parents are kept busy hunting for caterpillars and other insects on which to feed the chicks, although when adult the diet is mainly of small seeds picked up from the ground. Chaffinches are favourite garden birds, for they do no damage to crops and can become very tame and confiding when food is available from bird tables. In Britain they were at one time reckoned to be the most abundant kind of bird, for they were as common in towns and suburbs as in the country. Some years ago when the use of chemical pesticides and destruction of the

hedgerows was at its height, the numbers of Chaffinches began to drop and although they are by no means rare birds, they are not as common as formerly.

BRAMBLING —
Fringilla montifringilla

The Brambling like the Chaffinch to which it is closely related, is about 150 mm. long, but is a northern species, nesting only in Scandinavia and Northern Europe and Asia, but migrating to the rest of the Continent in the wintertime. This is when it may be seen in Britain, often in mixed flocks with Chaffinches and other small seed eating birds. These are usually to be seen in open country, but can also be seen in parks and gardens quite close to towns. In some years the Brambling population seems to increase dramatically and then the flocks may number many thousands of birds.

GREENFINCH —
Chloris chloris

Another very common visitor to parks and gardens where it feeds greedily on nuts and seeds provided at bird tables is the Greenfinch, which is a resident over much of Britain and much of Europe south of Scandinavia. Although only 145 mm. long, it gives the impression of being larger, for it is a plump, stocky bird. The male is far more brightly coloured than the female or young, which are a dull brownish green, although they both have the strongly forked tail and undulating flight characteristic of the species. The Greenfinch is an extremely vocal bird, for apart from the spring song of the male, which is a trilling mixture of notes and phrases, they have a wide range of calls. The nest is usually in thick cover and is built of grass, moss and roots. Here, several broods will be reared through the summer months and the fledglings continue to be cared for by their parents even after they have left the nest.

GOLDFINCH —
Carduelis carduelis

This elegant little Finch, measuring only 120 mm. long,

is resident in most of Europe south of the Gulf of Finland. The pleasing liquid twittering song and its bright colours have made it in the past, a favourite cage bird, and in some continental countries it is still to be seen in captivity, although this is quite illegal in Britain. Early settlers in America took their pet Finches with them and some were liberated; a small population of these birds survives in the north east of the United States. Unlike some other immigrants, such as Sparrows and Starlings, their numbers have never increased greatly and they are not in any way a pest species. Goldfinches are seen most frequently in places where well established vegetation gives them plenty of cover and in the wintertime particularly they are seen on the edges of fields, or waste ground where thistles and teasels have grown, for the seeds of these plants are among their favourite foods. The gardener may tempt them from hedgerows with cosmos which in Autumn may entice small flocks to feed on the seed heads. Although the territorial song of the male may announce its presence, in the summertime the Goldfinch is harder to see, for it prefers to remain in dense thick hedges and small trees. The nest, beautifully constructed of small twigs and roots, and lined with hair or wool is usually camouflaged with bits of moss or lichen and is usually on an outer branch of a tree.

SERIN —
Serinus serinus

The Serin which occurs in Britain only as a vagrant

species, is resident in Spain, southern France, Italy and the other countries bordering the Mediterranean. In the summer months it migrates as far north as the Gulf of Finland, but does not occur in Scandinavia or the Channel Coast of France and the Low Countries. This northern distribution is of relatively recent origin, and it is only since about the year 1800 that Serins have began to venture north of the Alps. They reached Holland in 1922 and bred in Denmark for the first time in 1949. It is also extending its range upwards

Chloris chloris

Serinus serinus

Carduelis carduelis

Hirundo rustica

in length, is not found in Britain but is resident over most of Europe, although it does not occur at high altitudes. It needs mature woodlands for its way of life and often finds the equivalent of this in well grown trees in parks and gardens. Here it may be seen climbing spirally up tree trunks, searching every crack with its long curved beak for small insects and spiders on which it feeds. Having exhausted the possibilities of one tree, it flies to the next, starting near the base and working quickly up. Many people catching a glance of a Tree Creeper at work think they have seen a mouse scuttling over the bark. The nest is usually hidden behind a piece of loose bark and both birds help with the incubation and rearing the young.

SWALLOW —
Hirundo rustica

The return of the Swallows from their winter home in Africa is a sign to the people of Britain and Northern Europe that Summer is on the way. These birds are to be found throughout almost the whole of Europe and Northern Asia and North America, where they are known as Barn Swallows. Barns, outhouses or porches may provide a site on a beam for the Swallow's nest, which is made of mud pellets and straw and lined with feathers. Here the Swallows may rear two broods of young before gathering in large, twittering flocks on telegraph wires or in reed beds. This is done in preparation for their long migration flight, in the case of the European birds to Africa south of the Sahara. The food of the Swallows is exclusively small insects, caught on the wing and the graceful, darting flight of the birds as they hunt, often over water, is one of the pleasures of the summertime for many people. The young birds, which have shorter tails than their parents, are not such adept fliers and have to be fed and cared for for some time.

HOUSE MARTIN —
Delichon urbica

Like the Swallow, the House Martin is a summer visitor from East and South Africa to almost the whole of Europe. It is a smaller bird, measuring only 125 mm. in length and although it also feeds on insects caught

in mountainous country, and is recorded as nesting at 1,350 metres above sea level. It rarely, however, lives in really isolated places, but seems to prefer the locality of houses and is even found in large towns, where parks, gardens or cemeteries offer it some degree of shelter. As well as this some birds remain in open forest areas, which is doubtless the original habitat of the species. As with most Finches, the nest is a deep cup made of fine roots and twigs and lined with plant fibres and wool and is the work of the female alone. It is usually situated on a horizontal branch close to the main trunk up to 6 metres above the ground. There are 3—5 eggs and incubation lasts about 13 days and as with almost all the Finches, is by the female alone, although the male remains nearby and feeds her on the nest. The young are fed by both parents on insects for up to 14 days and continue to be cared for for about another week after they are fully fledged. After breeding, Serins form small flocks, often with other seed eaters, and search for their food which consists largely of thistle, birch and alder seeds mainly on the ground. Some individuals remain in Central Europe right through the Winter, but most have migrated to the Mediterranean area by October. When they return in the next Spring, they often seek out the same spot in the same tree to build their nest.

SHORT TOED TREE CREEPER —
Certhia brachydactyla

This little bird, which measures only about 125 mm.

on the wing, it has a more fluttering flight. It breeds in colonies, usually under the eaves of buildings making a hemispherical nest of mud pellets, with a small entrance near the top. Two or sometimes three broods of 4 or 5 young may be reared before the Martins leave Europe in October. While Swallows are mainly associated with buildings in rural areas, House Martins often make use of houses in the suburbs and in towns for their nest sites, sometimes choosing buildings that have been standing for a few years only. In some areas, however, they may be at risk from sparrows, which if suitable nesting sites are in short supply, will drive the Martins from their half-built nest and take it over for their own nursery.

WAXWING —
Bombycilla garrulus

The Waxwing, which gets its name from the vivid coral-coloured tips at the ends of its secondary flight feathers, is a bird which breeds in the northern coniferous forests round the world. In wintertime it migrates southwards and may be seen in parks and gardens feeding on berries, especially those of the rowan trees, if these can be found. Measuring about 175 mm. long, Waxwings are obvious birds as a flock sitting in a tree, stripping the fruit entirely before moving to the next one to continue their meal. They seem to show little fear of man and to be extremely non-aggressive among themselves, keeping to fairly tight flocks in Winter and defending only the immediate vicinity of their nests in the summertime. Although Waxwings are

Certhia brachydactyla

Delichon urbica

Bombycilla garrulus

seen regularly as winter visitors on the eastern side of Britain, in some years huge numbers of birds move from their northern homes. These irregular irruptions carry them far beyond their normal range and they may then be met with as far south as the Mediterranean or as far west as Ireland.

GREAT TIT —
Parus major

With the exception of the far north, the Great Tit is resident throughout Europe. Originally a forest bird, it has adapted to man's presence and is now found commonly in parks and gardens, even in the middle of cities. Its bright colours, its confiding ways and its acrobatic antics as it feeds on nuts or other bird table goodies have made it a favourite with many town dwellers, who are further delighted by the fact that it readily accepts nest boxes in which to rear its young. Two large broods are reared, incubated by the female, but once the eggs have hatched, both parent birds are kept busy finding enough small caterpillars and other insects to satisfy the endless appetite of their family. In the late Summer, Great Tits form small flocks, often with other species of tits. They may stay near to houses where food is put out for them, but in more rural places may be seen in woodlands and scrub country, searching for small trees and bushes and even looking on the ground for seeds and grubs. They have a wide vocabulary of many calls, each apparently with a specific meaning. These have been intensively studied and have helped in our basic understanding of bird language.

BLUE TIT —
Parus caeruleus

The Blue Tit is resident throughout all of Britain and much of Europe, although it does not extend quite as far north as the Great Tit, which it resembles in its general habits although it feeds on the ground rather less frequently. Being smaller (about 112 mm. long), it tends to be more acrobatic. Ingenious experiments, designed to test its intelligence and co-ordination, have been made using wild birds coming freely to bird tables, where to obtain food they have to overcome various problems. The success with which these little birds cope with the strange situations with which they are presented is an indication of their intelligence and adaptability. One source of food which they have discovered in Britain is cream from the top of milk bottles, which they will peck open, quickly learning within a particular area the colour coding which distinguishes the creamiest milk. This habit is not widespread on the Continent, probably because milk is not bottled and sold as in Britain. Although they sometimes do damage in gardens by pecking at buds, their value as destroyers of large numbers of insect pests, especially in the summertime when they are rearing their large broods, makes them valuable friends of gardeners and horticulturalists.

LONG-TAILED TIT —
Aegithalos caudatus

The Long-tailed Tit is found over most of Europe except for the far north. In the most northerly part of its range it has, like the specimen illustrated here, a totally white head. In Britain and further south, a broad black stripe runs from the base of the beak, above the eye, and makes a partial collar where it meets the black of the shoulders. Intermediate forms are met with in Central Europe. Although it is 135 mm. long, about a third of this is taken up by the tail and the Long-tailed Tit is one of the smallest birds, weighing only 8—9 gm., to be seen in the countryside. It is extremely acrobatic, exploring the thinnest twigs for the tiny insects on which it feeds. It often enters gardens, but is not a common visitor to bird tables, preferring to find natural food in well-established shrubs and trees. Gardens often provide it with nest sites, for although it will not accept nest boxes, thick hedges, especially those of berberis or thorn are a favourite place for the rugby-football shaped nest, which is made of feathers, fine fibres and spiders' webs, decorated with pieces of lichen. Here, in April and May, the Long-tailed Tit rears its brood of up to 14 young. The parents are kept busy finding enough small insects for them, but often produce another family in June. Because of their small size, in very hard weather Long-tailed Tits suffer more than most birds. In the prolonged cold spell in the Winter of 1962—3, the majority of the population of this species in Britain died, the survivors, mainly birds in the south west, having subsequently recolonised the country.

BLACKBIRD —
Turdus merula

The commonest member of the Thrush family to be found in parks and gardens throughout Britain and most of Europe is the Blackbird. Only the male bird, which is illustrated, is black; the female is a dusky brown with a dark speckled breast. The strong, fluting song of the male, which is heard mainly at dawn and dusk, is thought by many people to be the finest European bird song. Blackbirds have adapted extremely well to living near to man. This may be seen in their choice of nesting places, for as well as nesting in trees, many birds choose to build in climbing plants, growing up the side of a house, or even such scanty

Parus major

Aegithalos caudatus

Turdus merula

Parus caeruleus

cover as a yard broom or a bundle of sticks leaning against a wall. The nest is a substantial affair, made with a base of twigs and lined with mud. Here, three broods may be raised during the summer months. Young Blackbirds continue to be cared for by their parents, even after they have left the nest. Well meaning people often try to 'rescue' fledglings, which appear abandoned, but the wisest course is to try and protect them from domestic cats and to leave them to the parents, which will certainly be about. The food of Blackbirds is very varied, and includes insects, grubs, worms, berries and fruits as well as food offered at bird tables.

SONG THRUSH —
Turdus philomelos

The Song Thrush is common as a resident in gardens and parks in Britain and much of Western Europe and is a summer migrant to most of the north of the Continent. The loud song of the male bird can be recognised by its habit of repeating each short phrase two or three times before going on to the next. The nest, which is usually built in a fork between the trunk and a branch of a small tree, is lined with mud and the young leave it well before they are independent, as with the Blackbird. Although it is the smallest of the common Thrushes (about 230 mm. long) the Song Thrush feeds on snails, a food source not usually tapped by other birds. In order to get at the soft body of the mollusc, the shell must be broken and the Song Thrush does this on a particular stone or anvil to which it carries any snail which it catches. Traces of broken shells round a stone are a sure sign of the work of a Song Thrush. Apart from this, it feeds on worms, grubs and fruit as well as bird table food. Although totally protected in Britain, members of the Thrush family may be trapped and used as human food in some European countries.

REDSTART —
Phoenicurus phoenicurus

The Redstart, which measures about 140 mm. long, is a summer migrant to almost the whole of Europe,

where it is to be found in open woodlands, parks and gardens. Both males and females may be recognised by the brick-red rump and tail, which is flicked up and down almost incessantly, although the female lacks the dramatic contrast of the male's black bib and grey head and back and is instead a warm brown colour. Redstarts seem to be constantly active and much of their food, which consists of small insects is caught on the wing in a flycatcher-like way. The song of the male is a pleasant, Robin-like twitter and the young also resemble immature Robins. The nest may be situated in a variety of places, including holes in walls, nestboxes or beams or ledges in buildings. The eggs are incubated by the female alone, but both parents care for the young once they have hatched. Two broods are normally raised in a year.

BLACK REDSTART —
Phoenicurus ochruros

The Black Redstart is a resident of much of Southern and Western Europe, including the southern counties of Britain and is a summer migrant to the more easterly parts of the Continent. About 140 mm. in length, it is like the previous species in many respects, although both male and female are darker in body colour. It is often easy to see, because when perching in bushes or on rocks, it usually chooses an exposed place to sit. Its original nest site was probably a ledge or cranny in a cliff face, but it has discovered that buildings make excellent substitutes for cliffs and it often uses them as such. During the 1939—45 war, Black Redstarts discovered that bombed buildings made ideal nesting sites and a colony became established in the City of London. Black Redstarts feed on insects, which they catch on the wing, but they spend more time on the ground than the Redstarts. Here they often move in a Wagtail-like manner.

CHIFFCHAFF —
Phylloscopus collybita

A characteristic sound of the summer months in wooded countryside or large overgrown gardens or parks over most of Europe is the monotonous song of

242

the Chiffchaff, which repeats its name over and over again. This tiny Warbler, which measures only 110 mm. long, is one of the earliest arrivals among the summer migrants and, indeed, in southern England may occasionally remain throughout the year, and is resident through most of the warmer areas of the Continent. It is extremely similar in appearance to its close relative the Willow Warbler (see page 50) although in general the colour of its legs is darker. Its preference is for breeding sites in dense cover, but above ground level, where five or six young are reared in a domed

nest. As with the other Warblers, only insect food is eaten.

GARDEN WARBLER —
Sylvia borin

The rather undistinguished looking Garden Warbler comes to Europe each Summer from Africa and may be found almost throughout the Continent apart from the extreme north and south. In Britain it is absent from northern Scotland, but elsewhere is common

Turdus philomelos

Phoenicurus ochruros

Phoenicurus phoenicurus

Sylvia borin

in the countryside and parks or gardens where there is dense undergrowth in which it can hide. Even the song of the male, which is a beautiful, mellow trill is delivered from thick cover. The male builds 'cock nests' which are the beginnings of nests, in rank vegetation in various parts of the territory. One of these is chosen and completed by both birds. Both help to incubate the eggs and to rear the young, which they feed on insects and their grubs, although later they eat soft berries of various kinds. At the end of the Summer Garden Warblers migrate to the savannahs and forest edges of Central and South Africa.

LESSER WHITETHROAT —
Sylvia curruca

The Lesser Whitethroat is a migrant from Africa to be found during the summer months over most of Europe, except for the hottest and coldest areas. It is very similar in size (about 135 mm. long) and general appearance to its close relative the Whitethroat (see page 173), although the dark ear patch is a good distinguishing feature. It tends to stay in cover more than the Whitethroat and is more often seen in trees, although the nest, which is built by both birds of a pair, is usually fairly close to the ground, in dense vegetation. When it has young it may become very noisy and protective and if they are threatened the female will feign injury to draw the attention of a predator away from the nestlings.

BLACKCAP —
Sylvia atricapilla

A brief glimpse will suffice to identify the Blackcap, for the top of the male's head is glossy black, while that of the female is a rich chestnut brown. It is similar in size (140 mm.) to the Garden Warbler to which it is closely related and it occupies similar habitats, although it has some preference for evergreen woods. It is a summer migrant through much of Central and Northern Europe, but remains through the winter in the countries bordering the Mediterranean and even does so rather rarely in Britain. After elaborate courtship and pre-mating behaviour, in which the male may perform a kind of dance, hopping up and

down on a branch, and chases with a special, slow flight, the nest is built by both members of a pair, usually in a bush or hedgerow. Incubation is by the female and during this time groups of males may assemble and posture. The best songs are heard early in the season, and are beautiful, varied warblings. The food of Blackcaps is mainly insects and spiders, but fruits of many kinds are also eaten.

SPOTTED FLYCATCHER —
Muscicapa striata

This little migrant, 140 mm. long, which arrives from Africa in April, is misleadingly named, for it gives the impression of being a dull brownish-grey bird, with very few spots. It is to be found throughout Britain and almost all of Europe, often in parks and gardens, close to mankind. Its rather untidy nest may be made in a variety of places, including creepers on the walls of houses. Here the female incubates and both parents rear the family of four or five chicks, sometimes producing two broods in a season. The Spotted Flycatcher may always be recognised by its habit of sitting very upright on an exposed perch from which it surveys the area for insects. When it sees a suitable one, it flies from the lookout, hawks after it and returns at once to the same spot.

GREY HEADED WOODPECKER —
Picus canus

The Grey Headed Woodpecker is a bird of Eastern Europe and has never been recorded in Britain. In many ways it looks like a small version (254 mm. long) of the Green Woodpecker (see page 51) but the red and black on the head is less extensive than in the Green Woodpecker and there is none in the female. In Central Europe, where the two species overlap, the Green Woodpecker tends to occupy more lowland woodland areas, while the Grey Headed is a bird of open forests of upland slopes, although it may also be found in parks or large gardens, even in towns. Like the Green Woodpecker the Grey Headed feeds on ants whenever possible and spends more of its time on the ground than other members of the family. The male has a laughing call, not unlike the yaffle of

244

the Green species, but in springtime drums with his beak on a dead tree to make a sound which carries through the woods to proclaim his territory. The nest hole is excavated by both members of a pair.

WRYNECK —
Jynx torquilla

This little bird (170 mm. long) gets its odd name be-

Sylvia curruca

Sylvia atricapilla

Muscicapa striata

Jynx torquilla

cause of its ability to cock its head at strange angles, especially if it is handled or disturbed on the nest, when it may also feign death. It is widespread in orchards and hedgerows through much of Europe and Northern Asia, which it visits as a summer migrant. In Britain, however, although it was moderately common during the last century, it has now become very rare and occurs in a few localities only, all of them in the southeast of the country. It is related to the Woodpeckers and like them may be seen searching the branches and trunks of the trees for insects and spiders, which it picks up with rapid movements of its long tongue, for it cannot excavate grubs from their tunnels in the wood. Ants are also an important food and it often searches for these on the ground. The Wryneck's voice is as strange as most other things about it; the main call is a shrill shriek, reminiscent of the smaller birds of prey, but it has a wide repetoire of sounds, which include snakelike hissing when it is disturbed on the nest.

SWIFT —
Apus apus

Swifts, which are often bracketed in people's minds with Swallows, are, like them, dark-coloured, long-winged summer migrants to almost the whole of Europe, where they may be seen wheeling about the sky catching the insects on which they feed. In fact, they are not at all closely related and both structurally and behaviourally the Swifts are quite distinct. Their length is only about 160 mm. but their long, scimitar-shaped wings span 385 mm. They are masters of the air, achieving higher speeds in level flight than any other species and spending more time on the wing including long periods at night than most. They never, unless injured or seriously weakened by lack of food, come to the ground, from which they find it difficult to take off. They fly directly into their nest holes and drop into the air as they leave. They breed socially in towers and old buildings, usually in towns and villages, making the nest from fragments, such as feathers, dead leaves or even bus tickets which they have caught in flight and which they glue to a beam or rafter with sticky saliva. When bad weather makes food difficult to find, Swifts may fly hundreds of kilometres. This

sort of journey means leaving the eggs — usually two or three in number — or the nestlings unattended. In most birds this would certainly mean their death, but in Swifts the eggs and young can cool and go unfed for several days without harm.

BARN OWL —
Tyto alba

The specimen illustrated is the Eastern European form of the Barn Owl. In Britain and the west of the Continent it has a silvery white breast, although the colouring of the back is similar in both cases. The species is native to most of Europe south of Scandinavia and has been taken by man to many other parts of the world. It is often, though not always, associated with man, for it tends to nest in old buildings, sometimes in towns, although hollow trees and holes in cliffs are also used. The Barn Owl hunts at dusk and at night, catching rodents, shrews and small birds as well as some insects. Its pale colour, silent buoyant flight and shrieking cry (it is known in some areas as the Screech Owl) have made it the subject of superstitions in some places. Although its eyesight enables it to see in very dim light, its hearing is so acute that it can detect its prey by using its ears alone.

LITTLE OWL —
Athene noctua

A number of colour forms have been described for the Little Owl, which, south of Scandinavia, is resident across Europe and Asia to North India. It was introduced into Britain during the last century and has established itself in the southern half of the country. It is generally a bird of open farmland, where it may sometimes be seen in broad daylight. It may be recognised by its small size (225 mm. long), its compact shape and relatively large head. It hunts mostly at dusk and dawn, feeding on rodents, small birds, other vertebrates and insects. It occasionally takes young game birds and for this reason was regarded as an enemy by gamekeepers, but in Britain it is now on the list of totally protected birds. If a large creature — a rat say — is captured, it will be taken to some secluded spot, such as a hollow tree to be dismembered. Food which cannot be eaten may be left in such a larder, but Little Owls do not hoard food systematically, although they may eat the insect larvae which are to be found in the carcasses rotting in their store. Little Owls make a wide range of sounds, including chattering cries and yelps and since these are often heard close to human habitation, these birds are held in superstitious awe in some parts of the Continent.

COLLARED DOVE —
Streptopelia decaocto

The Collared Dove was originally an inhabitant of Asia Minor and the Balkans but in the 1930s the numbers there began to increase and the species

Apus apus

Tyto alba

Streptopelia decaocto

Athene noctua

spread west and northwards. It is now resident from north Italy to Britain, where it first bred in 1955, and Scandinavia, where it is still spreading. 300 mm. long, it is very similar to the Turtle Dove (see page 61), but has an incomplete black and white collar round the back of the neck. It is often to be found near houses and smallholdings, especially where chickens are kept. It quickly becomes very tame and joins them at their food, but also eats some insects and molluscs. The nest is a flimsy affair, built on the horizontal branch of a tree. Here the female rears a series of broods, each consisting of only two squabs, but she often has young well into the Autumn. The voice of the Collared Dove includes a musical three-tone cooing, and rather harsh wailing calls.

MAMMALS

HEDGEHOG —
Erinaceus europaeus

The Hedgehog was originally an animal of deciduous forests in Europe and Western Asia, but has adapted well to life in more open countryside, as long as there is some cover in the form of hedges and copses. It is also found in parks and gardens, even in big towns, where it may be quite common and where it may become very tame, especially if encouraged by house-holders with bread and milk or table scraps. The body length of a large specimen may exceed 250 mm. and the weight may be more than 1,000 gm. for the Hedgehog is a plump, thickset animal, incapable of running fast on its short legs. Perhaps because of this it is nocturnal, hunting the slow moving insects, worms and slugs which are active then. It has no fear

of any enemies, for its back is densely clothed with sharp spines. If threatened in any way it rolls up into a ball, presenting an attacker with a deterring sphere of prickles. Unfortunately Hedgehogs react in this way to any danger, and they are the commonest mammalian casualties on the roads, for they have yet to learn that cars and lorries are enemies of a different sort from their natural foes. Hedgehogs carry an extensive fauna of parasites, but these are quite harmless to man or domestic animals. Unlike most insectivores, the Hedgehog hibernates in the wintertime.

MOLE —
Talpa europaea

The Mole is common, but rarely seen, because it spends most of its life underground. It is to be found through most of Britain apart from Ireland and in Europe occurs south of Scandinavia, apart from some of the Mediterranean countries, where it is replaced by a closely related species. The Mole's body, which may be up to 150 mm. in length is cylindrical, with little narrowing at the neck, and is clothed in dark, velvety fur. The limbs are short and the arms are twisted, so that the immense hands face outwards, as shown in the illustration. They are used to dig the tunnels in which the Mole lives and hunts. First it loosens the soil, then pushes with its snout and shoulders to consolidate the tunnel walls, although if the earth is too solid, molehills of spare material will be pushed up to the surface. The Mole has a living area or fortress, in which its nest is situated; here it may store surplus food to provide for the time when the weather is too wet or cold for digging. The Mole must eat nearly its own weight of worms and grubs every day and is a tireless hunter, but its activities

Talpa europaea

Erinaceus europaeus

often bring it into conflict with man, whose lawns and flower beds it may disfigure and whose farm machinery may be hampered by molehills.

WHITE TOOTHED SHREW —
Crocidura russula

This small insectivore may be found through most of southern Europe in parks, gardens scrubland and other open country where some cover is available. It is absent from Britain apart from some of the Channel Islands and does not occur in Denmark or Scandinavia. It is more strongly nocturnal than the Common Shrew (see page 64) which it resembles in size, and which is about 90 mm. in body length. The tail length, however, is variable, and there are some Mediterranean races with very long tails. The food consists of any small invertebrate which it can catch and this species enters houses more readily than most Shrews, to hunt the spiders and other small creatures which may be found there. A strange habit observed in this animal and its close relatives, is that if disturbed in the nest with well-grown young, the female will lead them to safety in a crocodile, each baby holding fast to the tail of the one ahead of it.

LESSER HORSESHOE BAT —
Rhinolophus hipposideros

Of the two major groups of bats to be found in Europe, the Horseshoe Bats may be distinguished by the complex flaps of skin on the face. Several species are present, of which this is the smallest, with a body length of about 40 mm. — quite considerably less than that of a mouse. It is the most widespread of the species, occurring from western Ireland, Wales and southern England southwards across Europe from Spain to the Balkans. It is in general a social animal and in the daytime groups may roost together in the darkness of lofts, cellars or caves. In the wintertime it is less likely to be found in buildings, but sleeps deeply in the depths of caves. Its food, which is caught entirely in flight at night, consists of various small beetles and moths. The flight of the Lesser Horseshoe Bat is rapid and erratic, with a fast wingbeat. When coming to land,

Myotis myotis

Rhinolophus hipposideros

it somersaults to catch hold of a projection of rock from which it hangs.

MOUSE — EARED BAT —
Myotis myotis

Bats are unpopular animals with most people. Their small size, their dark colour and their silent nocturnal flight, broken by piercing shrieks, which can be heard only by the young, all tend to make them animals of mystery, disliked by the majority. Superstitious beliefs, almost all untrue, have grown up about them, the commonest being that they will fly into women's hair, a situation abhorrent to any bat.

They are in fact a very successful group of mammals. Europe alone has 31 species and the world population is of about 800 species. All European bats are nocturnal or crepuscular insect eaters and are quite harmless to man. Bats are the only mammals to have developed a true flapping flight, the wings being formed of a web of skin which is stretched from the shoulders to the ankles and supported by the greatly elongated forearm, hand and finger bones. In some, but not all cases, this membrane continues from the ankle to include the tail.

Bat structure has had to be strongly modified to allow flight. While the forelimbs are huge, the hindlimbs are correspondingly small, but they play an important part in tensioning the trailing edge of the wing in flight and when at rest the bat hangs upside down by its toes. At such a time, the wings are puckered by elastic strands running through the membrane, which means that the animal can run over the ground or any other surface without damaging its means of flight.

Bats fly in swoops and ziz-zags so that although their flight is fast, it is less direct than that of most birds. European bats are not totally blind, but are extremely short-sighted and they navigate and detect their prey by means of sonar. In this, the bat gives off a series of high pitched squeaks, which quickly die away in the air, but if there is any obstruction, this will cause an echo, which is caught by the bat's supersensitive ears. As it approaches an obstacle, a bat is able to pinpoint its position and size and to take avoiding or pursuing action accordingly. Since flight is an activity requiring a great deal of energy, bats conserve this by becoming torpid when not fully active and by at least partial hibernation during the winter months.

There are two major groups of bats in Europe. One includes the Horseshoe Bats the other includes the Common or Smooth Faced Bats. The largest European species of these is the Mouse-eared Bat, which has a body and tail length of up to 140 mm and a wingspan of 370 mm. This is widespread over most of Europe south of Denmark, in the summer months, when it may be found in cellars, caves or lofts. As Winter approaches the northern populations migrate long distances to the south of the Continent. It has recently been proved to be resident in a very few localities in southern Britain.

SEROTINE —
Eptesicus serotinus

Another large species of bat is the Serotine, which has a body and tail length nearly as great as that of the Mouse-eared Bat, but shorter, broader wings. It is found in southern Britain and throughout Europe south of Scandinavia. It often shelters in old buildings, lofts or cellars, or in hollow trees in parks or on the edge of woodlands, from which it emerges to hunt early in the evening. It flies high, but with a more fluttering flight than another large bat, the Noctule, with which it might otherwise be confused.

LONG-EARED BAT —
Plecotus auritus

The Long-eared Bat may be immediately identified at close quarters by the size of its ears which exceed those of any other European species, although they are folded back when the animal is at rest. It is widespread through almost the whole of Europe south of Scandinavia, although it is absent from the most northerly parts of Scotland. It quite frequently takes shelter in houses or in trees in parks or gardens, but its presence may not be recognised because it does not normally fly until after dark. In flight it often appears to glide or hover close to foliage, from which it picks resting insects. In the Summer the females form nursery colonies, where the young are born; in Winter it may migrate to warmer areas.

GARDEN DORMOUSE —
Eliomys quercinus

This pretty little dormouse, which has an overall length of up to 320 mm. is to be found widely in Europe other than Britain and the far north of the Continent. It generally lives in woodland with a good shrub layer, but may occur in coniferous forest, or even among rocks in open country. The den may be in a tree, in which case it is likely to be an old bird's nest or squirrel's drey, or it may make a lair in holes in rocks or in a wall. It hibernates throughout the winter months, often sleeping from September to April and sometimes entering houses at this time, searching for a secure spot for its long sleep. When awake, it is active only at night, when it searches for its food which includes fruit and shoots, nuts and seeds, but more than any other Dormouse it will also eat insects, grubs, and molluscs and even young birds or mammals. It may do a good deal of damage in orchards and is said to store food in Autumn. Most Dormice are rather silent animals, but the Garden Dormouse is a highly vocal creature, making many types of growling, whistling and churring sounds. There are normally two litters of helpless young born to each female each year. These are quickly weaned and are independent in about four weeks.

HOUSE MOUSE —
Mus musculus

The House Mouse is probably native to the Mediterranean area and much of the steppe zone of Asia. Within this huge zone several subspecies occur and these are capable of cross breeding should they meet. In the south and east rather small, long tailed forms of House Mouse occur, and these are capable of surviving out of doors, beyond the influence of man. In the north and west, House Mice are usually larger in size — up to 100 mm. in body length and with a tail length roughly equal to this — and they are complete commensals of man, unable to exist without him. This mouse has been carried by man to every inhabited corner of Europe and beyond. Its high breeding rate — up to ten litters a year may be produced — ensures huge population increases wherever the con-

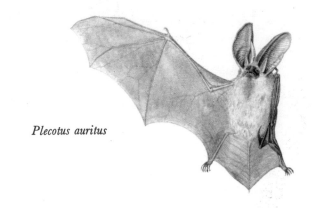

Plecotus auritus

ditions are favourable, although where populations are high, territorial behaviour develops and only the dominant male mates. The House Mouse is mainly nocturnal and feeds on a wide range of foods, often causing great damage to crops and stored products.

BROWN RAT —
Rattus norvegicus

Originating in North East Asia, the Brown Rat spread

Rattus rattus

Rattus norvegicus

Mus musculus

Eliomys quercinus

Martes foina

Mustela nivalis

Mustela ermina

Mustela putorius

to Europe in the early eighteenth century and is now found throughout the Continent and much of the rest of the world. With a body and tail length of up to 270 mm. and a weight which in a large specimen may exceed 500 gm. they are powerful, robust animals, capable of making extensive burrows, in which they live socially. Although not very good climbers, they swim well and are often to be found near water, for example in sewage systems, by rivers, or even along the sea shore. Their food is, for preference, grain, but they are quite omnivorous and will eat anything that offers. Many rubbish dumps have permanent rat populations, but crops, stored products and food of all sorts is taken by them. They often spoil far more than they eat, and they may damage buildings by gnawing beams or woodwork to gain access to food or shelter. They may spread disease, of which the most feared is weils disease — a form of jaundice. Their breeding rate is high, with several litters of up to twelve young produced in a single year by any female, so a rapid build up of large numbers can easily occur.

BLACK OR SHIP'S RAT —
Rattus rattus

In spite of its name, several colour forms of this animal exist, some of them quite pale in colour. It originated in tropical Asia and spread to the Near East in post-Roman times and is reputed to have been brought to Europe by Crusaders returning from their holy wars. It is now found throughout the warmer parts of the world, especially near ports, but has been largely supplanted by the Brown Rat in the colder areas, although it holds its own in houses, blocks of flats, shops or other large heated buildings. It is more elegant and agile than the Brown Rat and climbs extremely well but is a poor swimmer and it hardly ever makes burrows. Like the Brown Rat it is omnivorous and causes damage to grain and other stored products and it is also a reservoir of diseases, the most terrible of which is the bubonic plague. In suitably warm places it may breed throughout the whole year.

WOOD MOUSE —
Apodemus sylvaticus

This active little creature is extremely similar to the Yellow-necked Mouse (see page 67), but is slightly smaller, with a head and body length of up to 100 mm. and a tail which exceeds this. It is found widely through practically every habitat type in Western Europe, although in Britain it is rare above the tree line, although because of its fast breeding rate it may be very abundant in many other areas. Man's presence has not driven it from cultivated places, where it is usually a common inhabitant of hedgerows and it may enter houses, though it seems to do so less readily than the Yellow-necked Mouse. It makes quite deep burrows, especially if the soil is soft. Here the nest is made and stores are kept, for this animal is active through the winter months. In Spring and early Summer it may supplement its diet with insects.

POLECAT —
Mustela putorius

Two species of Polecat exist in Europe; the Western Polecat, *M. putorius*, which is the form found in Britain and the Steppe Polecat, *M. eversmanni*, which occurs in the South Eastern grassland areas of the Continent and across the Steppes of Asia. This species is pale in colour and may be the ancestor of the domestic ferret. The body length of the Western Polecat is up to 450 mm. and the tail adds another 160 mm. It is a ground living animal, choosing rough scrubland or rocky places, often near water to make its den, for although it is a poor climber, it burrows and swims well. It may often be found near to human settlements, where it is regarded as a mixed blessing, for as well as feeding on rabbits, rats and other rodents, it will, if the opportunity offers, attack chickens and game birds and for this it has been heavily persecuted. In Britain it used to be widespread, but it was brought to the edge of extinction during the latter part of the nineteenth century. However, a few survived in wild areas of central Wales, and from here they are now colonising south Wales. There are reports of their presence in Devon and the Lake District.

WEASEL —
Mustela nivalis

The Weasel is the smallest of the carnivores, and a small female may have a body length of only 60 mm. although their mates are much larger and a large male may measure 230 mm. The species is very widespread, being found from North Africa and the Mediterranean region throughout Europe (except for Ireland) eastwards across Northern Asia and also in North America. In Southern Europe, where there are

no Stoats, Weasels tend to be larger; in America the size is generally small. They are to be found in a wide variety of habitats, but prefer dry, sandy places where it is easy to excavate a den, which may also be made in the shelter of a heap of rocks, or among tree roots. They may be distinguished from their close relatives, the Stoats, by the lack of a black tip to the tail. They are mainly nocturnal, but may hunt by day also, preying on any small ground living animals, especially mice, voles and rats, which are killed by a bite at the base of the skull. Weasels are usually solitary, but the young may remain with their mother for some time after they might be independent, which accounts for stories of packs of Weasels.

BEECH MARTEN —
Martes foina

The Beech Marten is closely related to the Pine Marten (see page 69) but is slightly smaller in size and has a white, rather than a yellowish patch on the throat. It is a southern species, not found in Britain or Scandinavia, but it occurs in deciduous woodlands and the rocky southern areas of the continent. Although it is a good climber, it is less arboreal than the Pine Marten and it hunts more ground game including, in the south, a good proportion of reptiles and amphibians. It will also eat fruit, which forms an important part of its diet in the Summer and Autumn. Unlike the Pine Marten, which normally shuns human habitation, the Beech Marten is often found in the neighbourhood of houses, sometimes even in towns and it may make its den in a disused barn or stable.

STOAT —
Mustela ermina

This animal is absent from Southern Europe, but is widespread elsewhere in Britain, the Continent and across Asia and North America. It is considerably larger than the Weasel, with a body length of up to 290 mm. and a tail measuring about 100 mm. beyond this, although smaller specimens may be females or young animals. Stoats always have a pronounced black tip to the tail, which remains dark throughout the year, even in the coldest areas, where the animals turn white

253

in the wintertime. In the south this colour change does not occur and in intermediate areas, such as Britain, the change may not be complete. Any cover, whether woodland, hedge or rocky mountain slope will house a stoat, which is, however, rarely found as close to human dwellings as Beech Martens may be. Stoats feed chiefly on small rodents, hunting mainly at night, but they sometimes catch rabbits by day, fascinating them by leaping and somersaulting until they are close enough to dash. This is usually fatal to the rabbit.

IN
THE
MOUNTAINS

INTRODUCTION

Europe is a continent of many mountains. In the south we find representatives of the youngest mountain systems of the world, for the high peaks of the Alps, the Carpathians and the mountains of the Balkans are on a par with the high peaks of the Himalayas, the Rockies or the Andes. Further north, we find the worn down stumps of more ancient mountains and although these may give areas of rugged uplands as in the Black Forest, the Highlands of Scotland or the mountains of Norway they rarely show the jagged peaks characteristic of the younger folded-rock areas.

Mountains offer living space to many sorts of plants and animals, for within a relatively small space there may be differences of height, aspect and exposure which will make varied habitats. One feature which is always found in mountains is the cold, for the temperature falls about 1° C for each 500 metres in elevation. This means that the tops of high mountains may be permanently snow-covered, and nothing can live there, although some large animals, such as Ibex may take temporary refuge, and very small ones, like aphids or immature spiders, may be blown up there by high winds. These small creatures are frequently reported from extremely high levels, but their presence is accidental and they cannot hope to survive. Below the ice zone may be an area of steep, bare rock. This is attacked by rain, frost and wind and is most subject to erosion, which often leads to the formation of crevices in which hardy plants can take root. The edelweiss is one of the most famous of the extreme upland plants, but there are many others, frequently with very beautiful flowers. Lichens and mosses are also important in this rock zone. Below the rocks, comes a zone of grass and herbs, which may contain some bushes and dwarf trees. This zone is rich in life of many kinds and small mammals, birds and insects abound, while it is the grazing area of Ibex and Chamois. Below this come the forests of birch and conifers and finally, in some areas, broadleafed forests in the valleys. Here man has made most alteration to the environment and here villages and towns will be sited, for although there may be summer dwellings high on the mountain slopes, few are permanently occupied.

Conditions in mountains can change more rapidly than anywhere else. A rock face, baked by the sun, will lose its heat rapidly once the shadow of a neighbouring peak falls on it, or it may be rain-lashed one day and desert dry the next. As a result we find very great variety of life in close proximity in mountains and most animals which live there are either able to take cover and dig themselves in in adverse conditions, or they are sufficiently mobile to move quickly to more sheltered spot.

Mountains are one of the few areas that man has scarcely touched. As a result of this we find there many creatures which have been banished elsewhere. Eagles are found more plentifully in mountains than anywhere else, yet they could live well enough in lowlands were they not persecuted there. In Britain Wild Cats and Pine Martens are found only in mountainous areas and in Europe some species, such as the Ibex, would probably have followed the Aurochs and the Tarpan into extinction had it not been able to take refuge among the highest peaks. As man has as yet little economic use for the mountains, these are often the areas he is most prepared to set aside for nature reserves and game parks and in doing so often preserves the richest reservoirs of wild life.

Mountains, however, often have effects on life far beyond their boundaries. They are always areas of high rainfall, which may be in the form of snow in the wintertime. This acts as a reservoir for rivers, for when it melts in Spring, much of it rushes down in torrents which are the headstreams of many big waterways. Much more, however, remains in the upland bogs and this water will slowly become available as the Summer progresses. This will be used to regulate and modify the water supplies and life of the plains below.

Nucifraga caryocatactes

INSECTS

APOLLO BUTTERFLY —
Parnassius apollo

This handsome, large butterfly, with a wingspan of up to 80 mm. is absent from Britain, but is widespread in Europe, especially on mountain slopes in the south, where it may be seen at heights of up to 2,000 metres. Further north it occurs at lower altitudes, but is on the wing during July and August in both cases. The caterpillar is a spectacular animal, black in colour, with large red spots and blue warts on its sides. It may be found feeding on species of stonecrop during the early summer months, before it descends to the ground for its brief pupation.

AMPHIBIANS

ALPINE NEWT —
Triturus alpestris

The Alpine Newt lives in clear mountain streams at altitudes of up to 3,000 metres above sea level. About 80—120 mm. long, the male in his breeding dress is gaudy enough, having developed a low crest down the back, and a reddish colour, although at other times of the year he is not so bright. Especially in very high places, where the water is icy cold, the newt's activities seem to be slowed down, and egg laying may last until July, when the adults leave the water. The tadpoles hatch after 2—3 weeks, and may have along larval life ahead of them. In lowland areas they metamorphose to their adult form in a few months, but in some upland lakes this is delayed, often permanently. The tadpoles continue to grow in size and eventually become sexually mature.

BIRDS

NUTCRACKER —
Nucifraga caryocatactes

Nutcrackers are members of the Crow family found commonly in the coniferous forests of Central Europe, Southern Scandinavia and a band eastwards from there into Asia, where a subspecies with a thinner beak occurs. About 320 mm. long, its speckled plumage and, in flight, its rounded wings and the clear pattern of black and white on the tail should make identification easy. Nutcrackers feed on a wide variety of foods; insects and berries figure largely in their summer diet and during the Autumn and Winter they concentrate on various species of nuts, especially the seeds of *Pinus cembra*. Like jays, they make food stores, but instead of burying their treasures singly, they hide a number of nuts together in cracks in tree stumps or behind the lose bark of a dead tree. Usually they remember where their hoard is, so they can retrieve it later. Nutcrackers start breeding early in the year and by the end of February the substantial nest of twigs, lined with moss, grass and hair is already under construction in the dense cover of a conifer. Three or four eggs are laid and incubated mainly by the female. As with many northern species the population seems to fluctuate greatly. After a year of particularly suc-

Parnassius apollo

Triturus alpestris

cessful breeding, Nutcrackers, which normally merely descend to lower levels during the Winter, invade new areas, flocks often reaching as far west as Britain.

CHOUGH —
Pyrrhocorax pyrrhocorax

Another member of the Crow family, the Chough, is found in mountainous areas, mainly in Western and Southern Europe. It does not venture as high in the Alps and Caucasus as its cousin, the Alpine Chough, but is found mainly at altitudes of about 1,500 metres. It is also found round steep, rocky coasts, which is its principal habitat in Britain. It is the national bird of Cornwall, but few Choughs now remain on the cliffs of that county. Choughs are sociable and groups of 50 or 60 birds may breed together in a cave or an area where holes in the rocks give protection for the nests. They feed on insects and other invertebrates and also on seeds. They may become very tame and in many places they have discovered that mountaineers can be a source of food, so they are often found near to mountain-climbing huts. The Alpine Chough, which is found eastwards into Siberia, is recorded as having accompanied one of the Everest expeditions to a height of 9,000 metres.

RING OUSEL —
Turdus torquatus

The Ring Ousel is a migrant member of the Blackbird family, which is to be found during the summertime in the coastal zone of Norway, the upland areas of Britain and much of the highlands of Southern and Eastern Europe. It is about 240 mm. long and looks like a Blackbird (see page 241) although it is not quite so densely black and has a broad collar of white across the breast. The hen bird is paler and has the white patch less well marked. The Ring Ousel's song is like that of a Song Thrush, but with short chuckling sounds between the rather monotonous phrases. It is usually given from an exposed vantage point, such as a large boulder. The food is a mixture of small invertebrates and fruits and the birds generally prefer to be in a zone where there are at least scattered trees. The nest is usually built in one of these, although

it may be on the ground. The female incubates the eggs for most of the 14 days before they hatch, although the male feeds her and takes short turns on the nest. Two clutches of four young are normally reared before the birds return to the extreme south of Europe, Africa and South West Asia for the winter months.

REDPOLL —
Acanthis flammea

Although the Redpoll is a resident species throughout Britain, it is otherwise confined to the birch woods of the far north and the mountainous areas of South East Europe. As might be expected of an animal so patchily distributed through the Continent, a number of subspecies exist, the British form, for example, being smaller and darker coloured than that from Scandinavia. In the tundra zone it is replaced by a closely related species, the Arctic Redpoll. This little finch, which measures about 140 mm. in length, tends to be gregarious at all times of the year, even breeding in colonies, and a number of nests, in each of which 5 or 6 young are reared; these may be placed close together in small trees or bushes. In the Winter compact flocks search the ground for the seeds, but in Summer the young are fed mainly on insects or other small invertebrates. The twittering song may be given from a song post, or in flight. Redpolls from the north migrate, often in large flocks to Central Europe during the Winter. Here they often invade gardens along with Goldfinches and other seed eaters, among which they may be recognised by their speckled plumage and the bright red crown of the head.

WALL CREEPER —
Tichodroma muraria

The long, curved bill of the Wall Creeper enables it to winkle out insects and other small creatures from the crevices in rocks where they might be hiding. It is a bird of mountain heights, to be found up to the snow line in the Pyrenees, Alps, Carpathians and the uplands of Asia Minor, but it is nowhere common. In the Winter it may be driven to lower ground where it finds the conditions it needs for survival in walls and ruins, and where its fluttering flight, in which the bright colours of the rounded wings are shown, makes it easily recognisable. It is a solitary bird, apart from the breeding season, when a pair might be seen at the bulky nest, which is tucked into a rock crevice. Here the female incubates and rears 3—5 young, which are full grown at a length of 165 mm.

GOLDEN EAGLE —
Aquila chrysaetos

Europe's eagles are all becoming rarer than formerly, even the Golden Eagle, which is the commonest species. This great bird has a female which measures up to 910 mm. long, with a wingspan of nearly twice this. It is widespread from Western Europe across the Continent and temperate Asia, with the exception of the most extreme deserts, and occurs also over much of North America. It is the only Eagle species which still breeds in Britain, where it has very recently extended its range from its stronghold in the Scottish Highlands to the Lake District. Golden Eagles get their name from the golden sheen on the head and neck of the adults, which does not appear until their fourth year. They may be seen most frequently when soaring high over the ground, scanning for their prey. In Scotland this is mainly Blue Hares, Grouse or Ptarmigan and sometimes carrion lambs or deer. Eagles normally mate for life and at the start of the breeding season a pair of birds have a magnificent courtship display, in which they soar and dive, sometimes rolling over in mid air. The nest or eyrie is usually one of two or three, which may be used in rotation. Situated on an inaccessible crag, it may sometimes be built in quite a small tree, which, as the nest is added to each time it is used, may look scarcely strong enough to support the weight. Two eggs are normally laid. The chicks remain in the nest for nearly twelve weeks fed by both parents.

Turdus torquatus

Tichodroma muraria

Aquila chrysaetos

Alectoris graeca

Marmota marmota

ROCK PARTRIDGE —
Alectoris graeca

The Rock Partridge is a bird of the mountains of the Balkans and South East Europe, and it has been introduced into some areas of Central Europe where it is apparently thriving. Normally it is found at high altitudes and has been recorded as high as 3,500 metres in places where there are protecting bushes under

Rupicapra rupicapra

Capra ibex

which it can make its nest. Closely related species of the same size (about 330 mm. long) and similar appearance occur elsewhere in Europe, such as the Red-legged Partridge of the south west, which has been introduced into · Britain. Their general habits are similar, in that the female lays a large number of eggs which she incubates without any help from the cock, although he returns when the chicks hatch. These are capable of foraging for themselves within a short time of hatching, and feed on seeds, berries and leaves, but also eat such insects and small invertebrates as they can find. In spite of their independence their mother continues to protect and brood them for some time after hatching.

MAMMALS

ALPINE SHREW —
Sorex alpinus

The Alpine Shrew is an animal of coniferous forests and in Southern Europe lives in mountainous areas, although at its furthest north, in Germany, it is to be found at quite low altitudes. It is dark grey in colour and is a large animal as shrews go — about 75 mm. long in the head and body and has a very long tail which adds another 75 mm. to its length. It is interesting in that it is one of the few mammals found entirely in Europe and also that it overlaps in most of its range with several other species of Shrew. Its success in competing with them may be the result of several anatomical peculiarities not found in other shrews. Little is known of its life history, ecology or behaviour, so that such details as the use of the scent glands which are strongly developed on the sides of both sexes, is quite unknown.

SNOW VOLE —
Microtus nivalis

High mountains throughout the Continent are the home of the greyish brown Snow Vole, which is found up to the snow line in some parts of Switzerland, although it is commoner between 1,600 and 2,600 m. It lives in sunny areas where there is rock and grass above the tree line, digging tunnels with several entrances and exits in which it makes nest chambers and larders packed with dry grass and leaves. It is a large Vole, measuring up to 144 mm. plus the length of its tail which adds up to another 75 mm. The breeding season is shorter than that of most other Voles and the growth and development is slower than usual in these animals, although the first young born in any year probably breed before the end of the season. Their food includes all types of vegetation found at these altitudes but twigs of bilberry seem to be the favourite, if it is available.

MARMOT —
Marmota marmota

Another exclusively European species is the Alpine Marmot, which is found in open mountain country between 1,000 and 3,000 metres. Originally an Alpine animal, it has now been introduced into the Pyrenees and Carpathians. It is a plump animal with a body length of about 575 mm., a short tail and stumpy legs, but in spite of its unathletic appearance, it can disappear into its burrow with incredible speed at the least sign of danger. Its wariness is proverbial and it is said that at any time when Marmots are feeding, at least one member of the group is acting as sentry and guard, to warn the others with a shrill whistle should anything unusual appear. Marmots are social creatures, which live in small colonies, the position of which is probably dictated by the presence of enough soil for their burrows. These have several entrances and extend for a number of metres between blocks of stabilised scree, for the nest chamber is often more than a metre underground. The dry grass bedding is renewed in late Summer and in October the burrow entrances are closed and the whole group goes into the deep sleep of hibernation, which lasts until the next April. Breeding occurs soon after the spring emergence and the young, usually four in the litter, come above ground for the first time in July. They develop slowly, not breeding until their third year and then probably producing only one litter every two years. As might be surmised from this, Marmots are long-lived animals, and have survived up to twenty years in captivity.

CHAMOIS —
Rupicapra rupicapra

The Chamois is found in mountains from the Pyrenees to Asia Minor and has been introduced into certain other upland areas, such as the Black Forest and the Vosges. It is an easily recognised animal, goat-like in general appearance, with an oddly striped face. Its height at the shoulders is about 800 mm. and its head and body length is up to 1,300 mm. The most characteristic features are the curious horns, with the backward hook at the tip. There is some variation in size between the different populations, the western animals being smaller than those from the east of the Continent. Originally forest animals, and known from prehistoric sites at fairly low altitudes in the Alps, they now tend to inhabit the upper forests and beyond, roaming to the snow line in the summertime although they always retreat to lower ground in Winter. The females and young live in herds which unite to make groups several hundred strong in the rutting season, which is during November and December. These large groups break up early in the year and the lambs, born in May, are members of quite small herds. Chamois are active during the daytime, grazing and browsing whatever food is available in the area. They are extremely sure footed and can run and jump on dizzy slopes, although when moving in deep snow they will travel in line, each animal putting its feet in the tracks made by the ones ahead of it. Wolves and Lynxes will hunt Chamois when they can and Foxes and Eagles attack the young, but none of these is so damaging a predator as man, who has reduced the populations of Chamois very greatly in some areas. They have, however, suffered less than the Ibex, probably because they take readily to the cover of thick forests.

IBEX —
Capra ibex

Wild goats are to be found in many parts of Europe, but these are frequently the feral descendants of once domestic animals, as is the case with all the wild goats found in the remoter parts of Britain, for example. In the high mountains of Europe and Asia, at heights between 2,000 and 3,500 metres species of truly wild goat still exist. The European Ibex occurs in two forms; that from the Pyrenees is the smaller, standing a maximum of 760 mm. at the shoulder, while the Alpine form may reach 850 mm. Horn shape varies considerably, according to race, although the females always have smaller horns than the males. Ibex are creatures which normally live above the tree line. Incredibly sure footed, they can run and jump over precipitous slopes or loose screes, or walk along ledges which look too narrow to support them. They feed on shrubs, grasses and lichens and live in flocks, the males usually occupying higher ground than the females. The rut occurs in midwinter and the kids, which are usually born singly, but occasionally as twins, are born in May and June. Ibex have been hunted to extinction in much of their former range and even the Alpine race was reduced to a small handful of animals by the beginning of this century. As a result of careful protection the numbers have now increased somewhat, and Ibex have been reintroduced to some of their former haunts.

ON
THE
SEASHORE

INTRODUCTION

The long coastline of Europe is washed on the western side by the waters of the Atlantic Ocean which is split by the projecting land masses into the North Sea and the Baltic Sea. To the south, the Black Sea and the Mediterranean are almost enclosed areas of salt water, which have little connection with the great oceans of the world. Because of the influence of the North Atlantic Drift, which brings warm surface water from the tropics towards Europe, most of the coastline, even of Northern Europe, is ice-free for the greater part of the year. This may be contrasted with similar latitudes in North America, where ports are ice-bound for many months, because cold, Arctic waters creep down the coast, lowering the temperatures.

The form of the coastline depends on the rocks and the shape of the hinterland. Where, as over much of Britain and Western Europe, this is low lying, the coastline is gentle, with sandy beaches, or at best low cliffs. Where the land is mountainous, high cliffs may be formed where it meets the sea. Particularly in the glaciated areas of Scotland and Scandinavia, dramatic deep inlets, called fiords are formed. In other places, where great rivers run to the sea, extensive mud flats may be found.

The type of plant and animal life of the coast varies with the sort of shoreline. The least life is found on the extremely exposed coasts of some westward facing areas, where in times of storms, waves may crash hundreds of metres up the face of the cliffs, sweeping away all but the most firmly anchored of plants and animals. In slightly sheltered bays, or places where there is some protection from the battering force of the waves, life is abundant. Seaweeds, anchored by holdfasts to rocks and breakwaters, may offer shelter for the more delicate organisms, but many of the creatures of the rocky shore are protected by strong shells, which not only save them from the force of the waves, but also, at low tide, from the drying influence of the wind and sun. Limpets are perhaps the best example of this, for they can withstand battering from the waves and clamped tightly to the rocks are in no danger of dessication at low tide. Many fragile and beautiful creatures are found in this environment, but most of them live in the shelter of tiny promontaries or under overhangs which save them from destruction in time of storms.

Sandy shores look barren at first sight, for there is no place that the big seaweeds can find a hold. Under the sand, however, life is abundant. This includes many microscopic creatures, small enough to live in the water film around each grain and other, larger creatures, such as the Cockle or the Razor Shell. These creatures lie hidden below the surface, in most instances protected against the abrasion of any movement of the sand by heavy shells, so they are generally not highly mobile animals. They feed when the tide covers the sand, for then they push

268

up to the surface of the beach tubes, or siphons, through which they draw sea water into themselves. This contains oxygen and food, in the form of minute planktonic life, which is strained out using the same action for both feeding and breathing. The richness of the sandy shore may be gauged by the numbers of these molluscs, which may exceed 10,000 to a square metre. Their offspring are usually planktonic at first, and many may be eaten by filter feeders such as their own parents are, before the rest manage to drift to a suitable spot where they can settle down for the remainder of their lives.

A muddy shore is less attractive to most people than rocky or sandy coasts, but they are not less interesting and are as filled with life as are the other coastal habitats. Many kinds of worms, molluscs and other invertebrates are to be found in the mud, and the richness of this fauna tempts birds to the mud flats where they can find abundant food. Muddy shores are often characterised by salt marsh flora. This is composed of flowering plants, related closely in some cases to plants of the land, but here adapted to withstand inundation in salt water. The roots of the plants help to consolidate the mud and in some cases the stems and leaves trap silt, often causing rapid growth of the marsh and ultimately the dry land at the expense of the sea.

Vertebrates are found in all sorts of coastal environments. Many kinds of fishes are to be found in the rock pools of the exposed coastline, some of them, like the Lumpsuckers, adapted to attach themselves to rocks so as not to be swept away in time of storms. Others, like the Weever, may bury themselves in the sand, while others such as the Sea Horse, may be able to attach themselves to the seaweed offshore and prevent themselves from being carried out to sea, where they could not hope to survive. Sea birds have to come to land to breed, although this is the only time that many of them set foot on the land. They mostly nest in dense, noisy, smelly colonies, some on cliffs, others, such as some Gulls and Terns, in broken land at the top of cliffs or even on sandy or pebbly beaches. Generally, however, they are only temporary inhabitants, and many of them, when their young are reared, make for the open sea, where they remain until the next Spring. A few birds, such as Kingfishers or Herons, may make the seashore their home for a short time, but they are really birds of fresh water, to which they always return. Mammals are represented by Seals, Porpoises and Dolphins. The Seals need to come ashore to rest and to produce their young. At this time they are vulnerable to human hunters and in many places their numbers have been greatly reduced from their former abundance. In the Mediterranean both Seals and Porpoises are now extremely rare, although they are common in some northern areas. The Polar Bear is an inhabitant of the far north and sometimes gets carried far south on ice floes, or may even swim for great distances and has been recorded as reaching as far south as Scotland, although not in recent years. This animal, now widely protected, has also been much reduced by indiscriminate hunting.

MOLLUSCS

OYSTER —
Ostrea edulis

The Oyster is well known to many people who have never so much as seen the oceans, for this mollusc is one of the favourite delicacies to be harvested from the shallow sea. Many species are known from all parts of the world, but the most prized for its flavour is the Native Oyster, which is found round the coasts of Britain and the adjacent continental shores. Its place is taken to the south by the Portuguese Oyster, but this requires higher temperatures for breeding than are reached in British waters, so while it is often grown in the north, it does not reproduce or establish itself there. The adult Oyster is entirely sedentary, for it cements itself by its left valve to a suitable part of the sea bed. The right valve then becomes a sort of lid to close the shell, which grows unequally in an irregular round or oval shape. The shell is thick and marked by uneven growth lines and may be up to 100 mm. in diameter. When thrown up on the beach, Oyster shells may often be seen to have been bored into by a sponge, *Cliona ciliata*, which honeycombs the thick calcareous valves with its activities. In the course of its life any individual oyster is capable of changing from female to male, but is never capable of fertilising its own eggs. Huge numbers of these are shed into the sea when the water temperature reaches about 15 °C. and breeding will continue so long as this is maintained. A single individual may spawn as many as 3,000,000 eggs in a season, but few of these can hope to survive. They are at first planktonic but finally settle, usually in an estuary or creek which does not dry out at low tide. The Oyster was certainly a major food of some prehistoric men in Western Europe, but the Romans were the first to bring any systematic harvesting to the Oyster beds, for they considered the flavour superior to that of the southern species and they exported huge numbers from the north of their empire to Rome, where vast quantities were consumed. The demand dropped during the Dark Ages, but from Tudor times on, the popularity of the Oyster increased. Even so, the stocks remained high until the present century, when the combination of over-fishing, disease and pollution decimated their numbers. Natural enemies, such as Starfishes, which can pull the shells apart with their sucker-ended tube feet, continue to be a pest, but they have been joined by various imported species. These were brought in accidentally with stocks of Portuguese and American Oysters and reared in the traditional beds. The new enemies include the Oyster Drill, which can bore through the shell to get at the succulent flesh of the mollusc, the Slipper Limpet, which smothers the Oysters and prevents them from feeding and an introduced Barnacle, which takes a heavy toll of the young, or spat. Oyster farming has been practised in some parts of the world for a considerable time. At its simplest it merely involves the spreading of suitable, clean, shelly material on which the spat may settle, and from which they can get sufficient calcareous matter to build their shells. More advanced methods involve breeding the Oysters in tanks and supplying the spat with adequate living space for very large numbers to settle within a relatively small area. Semi-circular tiles, attached to piles driven into the sea bed offer excellent attachment areas, but these must be placed in the nursery at exactly the right time, for the spat must be present, yet the tiles must not have become silted. When they are two years old, the young Oysters are removed from the nursery and put into boxes with wire mesh lids, which will enable them to continue to feed on plankton, but which is fine enough to exclude their enemies. Later they are moved again and this time left undisturbed for three years growth. Often, before they are finally harvested, they are put into an area of particularly rich food, for a first fattening session. The numbers of Oysters eaten is still enormous, but demand exceeds supply, even at the high prices which they command at the moment, and in spite of strict regulations regarding the time at which they may be gathered. The supplies of unfarmed Oysters show no sign of returning to their former abundance and although several other species of Oyster exist, these are not generally eaten.

COMMON PIDDOCK —
Pholas dactylus

This strangely shaped bivalve can never close its

273

shell completely, like the Cockle or most of its relatives. The reason for this is that the Piddock bores into soft rocks such as chalk, some sandstones, shales and also into wood. It does this by extruding the muscular foot through the gape hanging on to the rock with it while it rocks the shell, turning slightly as it does so. The sharp, tooth-like leading edges of the shell cut a tube in the substrate, in which the Piddock is protected against all enemies. It is found on the lower shore from the southern coast of Britain to North Africa. A number of species of Piddock exist, of which this is the largest found in Britain, reaching a length of 80 mm. Because it does not normally bore into wood, and cannot tackle the hard stone or concrete of which sea walls and quays are made, the Piddock, unlike its cousin the Shipworm, does relatively little harm to man.

COCKLE —
Cardium edule

The Edible Cockle is one of a number of closely related species which live in sand, mud or fine gravel between the tide lines or just off shore. It has a wide range, for it is found from the Barents Sea and the Baltic to the Mediterranean and round the coast of Africa to Senegal and it is also known from the Black and Caspian seas. It is a plankton feeder and where supplies of such food is rich, may grow to a size of 50 mm. or more in length, but this is unusually large. All of the Cockle species have a heavy, rather rounded shell, which makes it difficult for them to burrow fast or far and they normally live only a few centimetres below the surface of the sand. If an Edible Cockle is dug up, however, it will quickly rebury itself using its muscular foot, which is pushed into the sand dragging the shell after it. It is strongly angled and this enables the creature to hop along over the surface for a short distance, but in general they are sedentary animals. In France and the Mediterranean countries several species of Cockles are important sea-foods, but in Britain only this one species is normally eaten. The heavy shell, buried beneath the sand protects it from most enemies other than man, one other is the Oystercatcher, which probes in the sand and levers the valves of the shell open with its long beak.

LIMPET —
Patella vulgata

The first animal to catch the eye of the visitor to the sea shore is likely to be a Limpet. These creatures are in fact snails, but with a conical rather than a whorled shell. Tightly attached by a strong, muscular foot to rocks or breakwaters, they scarcely seem to be alive, but when covered by the sea at high tide, they leave their homes and may travel as much as 25 cms., rasping with their file-like tongues at any seaweed growth on the rocks as they go. In places where the substrate is soft the tooth marks of the Limpets may be seen over their grazing grounds. As the tide falls, they return to their attachment place, where shell and rock have been ground together to make a perfect fit to prevent drying out at low water.

SCALLOP —
Chlamys varia

A number of kinds of Scallops occur off the coasts of Europe, their unequal bivalve shell being among the easiest to recognise among the molluscan debris to be found on the beach. The specimen illustrated is one of the smaller species, rarely growing much above 60 mm. in length. Others, such as the Great Scallop, may be over 150 mm. in length. This and the smaller Queen Scallop are both fished commercially. Most Scallops are less immobile than bivalves are in general, for they usually live unattached to anything on the sea bed and are capable of moving rapidly if danger threatens. This they do by clapping the two halves of the shell together, expelling any water which may be there. Jet propelled in this manner, they hop clumsily away from enemies such as Starfishes, which they can detect not only by their sense of smell, but also by sight. The living Scallop has a number of pearly eyes placed between the tentacles on the edge of the mantle, just inside the shell.

COMMON MUSSEL —
Mytilus edulis

The Common Mussel is one of the most widespread species of molluscs, for it is found everywhere south of the Arctic to the Mediterranean and on the east coast of the United States and in Japan. Very similar species are found elsewhere in the world. Their purplish blue shells, which may grow to a length of 150 mm. (the largest recorded is over 200 mm.) are easily recognised and where conditions are suitable, they may be extremely abundant. They live from the middle shore to below the tide line, in dense colonies attached to rocks and breakwaters by a bundle of fine but strong threads called the byssus. The largest and most commercially valuable specimens live below in the sub-littoral zone where there is a constant supply of the plankton and suspended matter on which they feed. The intertidal beds, which never carry such large specimens, are nonetheless valuable as a source of Mussels for fattening in specially prepared pens. Because they are filter feeders, Mussels often take sewage particles and other materials which while not immediately poisonous to them, may make their flesh unsuitable for human consumption. Before being ready for the market, they have to go through a period of cleansing, without which they could be dangerous to eat. In areas where the amount of grit, etc., suspended in the water is high, Mussels often get particles inside the shell, and may form pearls to surround the source of the irritation.

CRUSTACEANS

ACORN BARNACLE —
Balanus balanoides

Several species of Barnacle, all rather similar to the one figured, occur on European beaches. They are separated by details of their structure and life histories, but are known collectively as Acorn Barnacles. They are among the most abundant animals on any rocky shoreline, but are not, as their limy shells suggest at first, molluscs, but crustaceans, related to the Crabs and Lobsters. This is never apparent in the adults, but the larvae of the two groups are very similar in their structure and development. Barnacles are hermaphrodite, but each fertilises the eggs of a near neighbour. The larvae when released into the sea — usually in early Spring — are minute spined creatures which

float in the surface waters, feeding on organisms even smaller than themselves. They grow and moult into a form which finally choses a piece of rock on which to settle for the rest of its days. Some shade, and a rough surface seem to be important factors in making the decision, and a Barnacle examines a potential home area very carefully for some time before settling, head downwards, on the rock. It then develops a series of limy protective plates which can be closed against enemies and dessication at low tide. When the

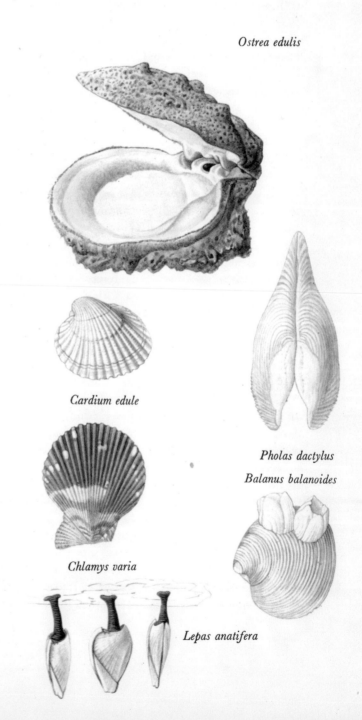

Ostrea edulis

Cardium edule

Chlamys varia

Pholas dactylus

Balanus balanoides

Lepas anatifera

sea covers it, it extrudes six pairs of feathery appendages and combs the water for food with a quick, grabbing movement, obtaining oxygen at the same time. In general, those individuals living lower on the beach grow more quickly, but live for a shorter period than those nearer the high tide mark. The numbers in either place, which may exceed 30,000 to a square metre, indicate their efficiency as utilisers of the sea-shore environment.

GOOSE BARNACLE —
Lepas anatifera

Goose Barnacles are animals of the open sea, where they attach themselves by a fleshy stalk to floating wood or other debris and like Acorn Barnacles kick food into their mouths with their feather-like feet. They are much larger than Acorn Barnacles, reaching a shell length of 20 mm. and are much more fragile in appearance, for the stout limy plates of the shore species are replaced with brittle but transluscent protective covers. For many years the exact relationships of these animals was in doubt and in medieval and later times the myth grew up that Goose Barnacles grew on trees and changed into geese, notably the species known as the Barnacle Goose. The manifest nonsense of this was doubtless perpetuated by those who, counting Barnacles as fish, thus felt able, with a clear conscience, to eat goose on Fridays and during Lent.

SEA SLATER —
Ligia oceanica

This creature, which is widespread round the coasts of Europe at about high tide level is a relative of the land Woodlice (see page 215). It is a general scavenger, up to about 25 mm. long and is to be found in cracks in rocks, or on sea walls or under seaweeds. If disturbed it can run fast to a new place of safety. Because its relatively large size makes it obvious to predators, the Sea Slater is active mainly at night.

SKELETON SHRIMP —
Caprella linearis

Among the strangest of animals of the sea shore are the Skeleton Shrimps. These little creatures, which are less than 12 mm. long must be searched for among seaweeds on the lower shore, but they will not be easy to find, for their reddish coloured bodies merge into the background of the red weeds. Slender as a piece of thread, they hang on with their hindlimbs, while the front ones are held up almost like those of a Praying Mantis. It moves slowly and deliberately, feeding on detritus and any small animals which may come within its grasp.

LOBSTER —
Homarus vulgaris

The living Lobster is blue in colour, not pink, which is the colour it goes when it is cooked. It is normally to be found under overhangs in rock pools on the lower shore in the summer months, but these are likely to be small individuals; the larger animals, which may measure up to 500 mm. in length are sub-littoral and are fished for using Lobster pots, or by skin divers in offshore waters. The Lobster can defend itself against most enemies with its powerful claws. These it can shed, without harm to itself if one gets caught in a crevice or is too firmly grasped by a predator. The lost claw begins to regrow with the next moult and often Lobsters with claws of markedly different sizes may be found. Another method of defence is flight and in spite of its weight the Lobster can, with a flick of its strong tail, shoot backwards out of trouble. In spite of these defences the Lobster, which has for long been regarded as a delicacy, is helpless against human fishers and the species is now in a decline from the North Atlantic to the Mediterranean. The females, which carry their eggs attached to the abdomen, are supposedly protected when in this state, but many are taken and the populations dwindle accordingly. Unfortunately, it has not yet proved practicable to farm Lobsters for once the eggs have hatched, they lead a complex planktonic life and how to provide the correct food in the right quantity for them is a problem as yet unsolved. The adult Lobster feeds on a variety of dead and decaying animal matter. One claw is usually used for crushing and the other for cutting its food, so that no carrion is too large, or too tough for the Lobster to tackle.

PRAWN —
Leander squilla

Prawns may be seen in rock pools during the summer months when they can be watched as they search the water for any food particles, using their two pairs of antennae and their stalked eyes to discover fragments of plant or animal debris which they pick up delicately with the tiny pincers on the first two pairs of legs. During the winter time these 50 mm. long crustaceans depart for offshore waters, where they will be warmer and better protected against storms.

SHRIMP —
Crangon vulgaris

Shrimps are to be found in pools or estuaries with a sandy bed and are so hardy that they may occur equally in northern waters where the temperature is low, or in the south where it may reach 30 °C. They can also stand low salinities, so the brackish conditions found where rivers meet the sea are no bar to them. Shrimps may grow to a length of 80 mm. but even at this length are difficult to see, for their bodies are almost transparent. For further protection they may bury themselves in the sand during the daytime, becoming active and feeding on a wide variety of plants and animals at night. As with Lobsters and Prawns, the females carry the eggs attached to the abdomen and may spawn twice in the course of a year. They are the basis of important fisheries in many areas, such as Morecombe Bay, round the coast of Europe.

HERMIT CRAB —
Eupagurus bernhardus

This animal, which is common and widespread round the coasts of Europe is Lobster-like, in spite of its name, and has a long tail. This, however, is unarmoured, so to protect itself the Hermit tucks it into the empty shell of a dead sea snail. The opening of the shell is closed, when the Crab withdraws for safety, by its one large claw, for in order to fit the spiral of its home it is asymmetrical in its general structure. Small Hermits may be found on the shore inhabiting top or periwinkle shells. Larger specimens live offshore, normally

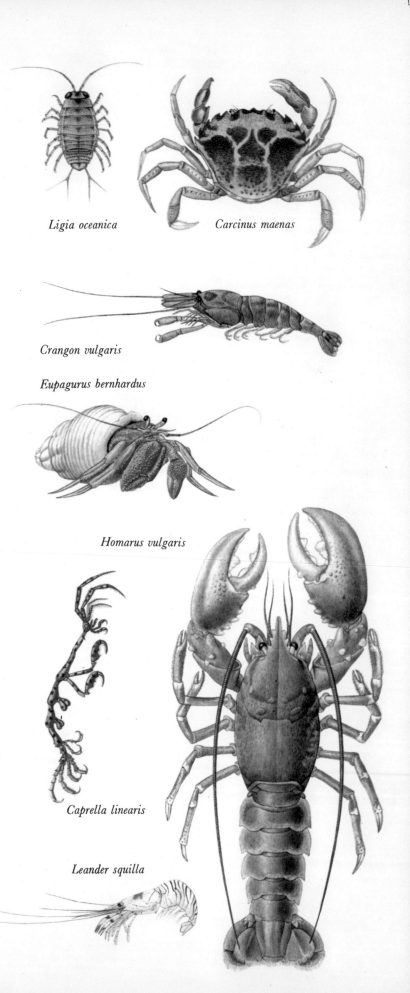

Ligia oceanica

Carcinus maenas

Crangon vulgaris

Eupagurus bernhardus

Homarus vulgaris

Caprella linearis

Leander squilla

Macropodia rostrata

in whelk shells. They are general scavengers, but may share their food with a number of commensals, which live in and on the shell. These include hydroids, sea anemones and a kind of ragworm.

SHORE CRAB —
Carcinus maenas

This is the commonest species of Crab round the coasts of Britain and much of Europe. Usually greenish in colour and growing to a shell size of 100 mm. across the shell it is to be found on the middle and lower shore of almost all kinds of beach. It can burrow into sandy spots or take shelter in cracks, in stones or under rocks, but will always be ready to defend itself with its sharp claws or to tackle any kind of food, mainly dead or dying animals. In the summertime the females carry large numbers of tiny eggs tucked under the abdomen. These hatch into planktonic larvae which must spend some time floating in the sea before settling down to an adult life.

SPIDER CRAB —
Macropodia rostrata

The many kinds of Spider Crab may be recognised as such by their small bodies and extremely long legs. This species, which is very common under stones and seaweeds on the lower shore is usually reddish in colour and has a shell span of about 12 mm. Many Spider Crabs, which are small-scale scavengers, protect themselves against enemies by disguising themselves with scraps of seaweeds or sedentary animals, which they place on their shells.

FISHES

LESSER SPOTTED DOGFISH —
Scyliorhinus caniculus

This small member of the Shark family, which reaches

a maximum length of 760 mm., is common round the coasts of Europe. It is a bottom living fish and feeds mainly on molluscs and crustaceans, although it may occasionally catch some of the sluggish fishes of the sea bed, including particularly various species of Goby. It is an important food fish, usually sold under the name of 'Rock Salmon', its true nature disguised by being skinned and beheaded. In the Winter the females lay a small number of eggs, each in a quadrangular horny case, about 50 mm. across, with a long, curled tendril arising from each corner. These are entwined in fronds of seaweeds, where the little Dogfish develop. Later the cases may be cast up on the beach, where they are known as 'Mermaids' Purses'.

THORNBACK RAY —
Raja clavata

Rays and Skates are also related to the Sharks, but they almost all live on the sea bed, where their flattened shape helps them to escape detection. The Thornback, so called because of the spines on the back and tail, is widely distributed in European offshore waters and also occurs in the southern Baltic and the Black Sea. Feeding on bottom living invertebrates and fishes, it may grow to a body length of 850 mm. and a weight of over 17 kgms. As with the Dogfish, the female lays her eggs singly in horny cases, which are about 75 mm. across. These have points, rather than tendrils at the corners, which is a characteristic of all Ray egg cases. This type of Mermaids' Purse may be easily recognised as one side is more convex than the other.

ALLIS SHAD —
Alosa alosa

The Allis Shad, which is a member of the Herring family, is to be found from the Gulf of Finland southwards round the coasts of Europe, Asia Minor and North Africa. It lives in shoals in fairly shallow waters and in early Summer the adult fishes, which may measure up to 700 mm. in length, gather for their migration up river, for like the Salmon, the Allis Shad spawns in fresh water. Although it stays mainly in the estuaries, it has been recorded as much as 800 km. upstream. The eggs are shed in areas of clean sand or gravel and may

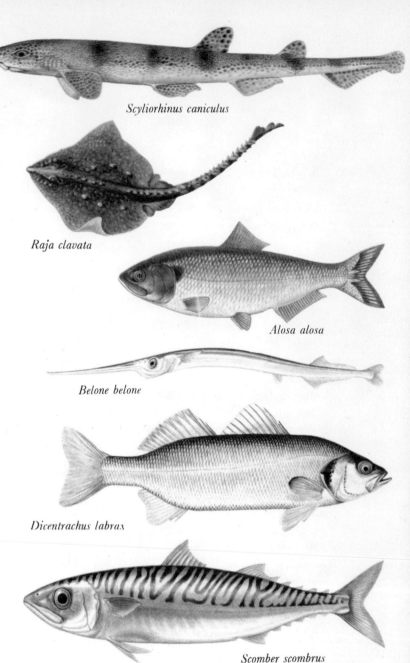

Scyliorhinus caniculus

Raja clavata

Alosa alosa

Belone belone

Dicentrachus labrax

Scomber scombrus

float downstream somewhat before they hatch, which is usually in under a week. Later in the year, the young migrate to the sea, where they grow rapidly, feeding on small shrimps and other crustaceans. Pollution and the blocking of rivers with dams and waterworks has reduced the chances of survival of the Shad, which is now quite a rare fish in Europe.

GARFISH —
Belone belone

The elegant, long bodied Garfish is an inhabitant of the surface coastal waters round Europe and in the Black Sea and related species occur elsewhere in the world. It has been recorded as growing to a length of 760 mm., but is usually smaller than this. It feeds on many kinds of small fishes and crustaceans among the larger plankton, which it snaps up in its long jaws as it skims along in the upper waters. Spawning takes place in shallow water in early Summer and the eggs, which have long, sticky threads attached to them, become entangled in the offshore seaweeds. The jaws of the young Garfish grow at an uneven rate and at one stage in their development there is even more discrepancy between the lengths of the upper and lower jaws than in the adults. Garfishes often accompany shoals of Mackerel, and may be caught in the nets set for them. Although they make excellent eating, they are not popular table fish probably because the bones are a brilliant blue-green in colour, which many people find distasteful.

BROAD NOSED PIPE FISH —
Syngnathus typhle

This fish is found from the Baltic to the British coast, where it is widespread but uncommon. It may be recognised by its size, which reaches a length of 300 mm., although it may be mature at half that length, and the rather broad snout on the sea-horse like head. It most often occurs in dense beds of eelgrass where it lies almost motionless sucking in the many sorts of small organisms, which include small crustaceans and fish fry, especially Gobies. Spawning takes place from June to August when the female lays her eggs into a pouch, which can be seen in the illustration

on the underside of the male. Here they develop and even after their release into the water, the young fish return to the pouch for safety.

BASS —
Dicentrachus labrax

The Bass is an inhabitant of European coastal waters from the Mediterranean to Britain, although since

Syngnathus typhle

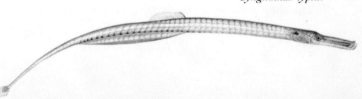

Lophius piscatorius

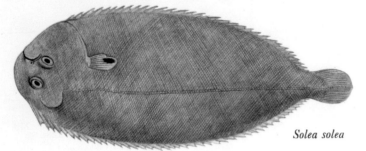

Solea solea

it is a warm water fish, it is rarely caught further north than Wales, and disappears during the winter months. It is not known exactly where it goes at this time and it has been suggested that it hibernates in deeper water offshore. It may grow to a length of 1,000 mm. and a weight of 8—9 kgms., but it seems to be a very slow growing fish and may take 20 years to reach a size of 750 mm. Spawning takes place inshore, often in estuaries, between April and June in British waters, but starting as early as February further south. It feeds on small fish and in the south follows the shoals of sardines. They are themselves excellent eating and are highly prized by inshore anglers.

MACKEREL —
Scomber scombrus

Mackerel are streamlined, fast swimming fishes, often to be found in vast shoals round Britain and in European waters. The British populations spawn from about April to June, over rather deep waters near the edge of the Continental Shelf but subsequently move towards the shore. Later still in the year, they congregate in certain localised areas of the sea bed, but move away from these as the breeding season approaches again. Mackerel are mature at the age of two years, when they measure about 300 mm. long, but they

are long lived fishes and continue to grow until they may reach 500 mm. Their food varies according to the time of the year. After spawning they feed chiefly on small crustaceans, which it is said that they hunt individually, rather than straining them from the water. Later they feed mainly on young fishes, which they often follow close in to the shore. They are caught in great numbers, both by nesting and on lines, for their flesh is excellent food, in spite of old wives' tales to the contrary. These tales are probably based on the speed with which the beautiful iridescence of the Mackerel fades after death.

DOVER SOLE —
Solea solea

The Dover Sole is one of the large group of fishes generally known as 'flatfish'. They all lead an inactive life on the sea bed and as their name suggests, the body is flattened. This is quite different from the flattening of the Skates and Rays (see page 278), which are flat from above and are symmetrical animals. In the Sole and its relatives, the flattening is lateral and the fish, which starts life as a normal shaped creature, lies on one side as it develops — in the Sole it is always the left side — and the eye from this side swivels round until it is in a lopsided position on the top (right hand side) of the body. The mouth is also asymmetrical, and when feeding, which it does mainly at night, the Sole moves slowly over the sea bed, examining it with special touch organs on the underside of the head. Any small worm or mollusc perceived in this way will be pounced on, but suspended food cannot be found. The Dover Sole, which used to be fairly common round the coasts of Britain and Western Europe has now been heavily overfished and large specimens, up to 500 mm. in length, are now comparatively rare.

ANGLER FISH —
Lophius piscatorius

Another slightly flattened bottom living fish is the Angler, sometimes called the Monk Fish. It is found from the North Sea to the Mediterranean in fairly shallow water, where it may grow to a size of over

1,500 mm. It is related to a number of deep sea species, with which it shares the characteristic of having a long, flexible spine, derived in fact from the original support of the dorsal fin, on the top of the head. This has a small flap of skin on the end of it which acts as a bait to lure unsuspecting small fish within reach of the Angler's outsize jaws. Observations on Anglers have shown that the 'bait' is never touched by the victim, but is jerked about in front of it until it is engulfed. Many small fishes and crustaceans fall prey to this trick, but in spite of the size of its mouth, the Angler is not capable, as are some of its deep water relatives, of swallowing fishes of a greater bulk than itself. It is disguised by its colour, which merges with that of the sea bed and also by having a flap of skin round it, which breaks up its outline. Anglers are widely eaten in Southern Europe, but are rarely offered for sale in Britain and then only when skinned and filleted.

BIRDS

WHITE TAILED EAGLE —
Haliaeetus albicilla

This bird, sometimes known as the Sea Eagle, is generally to be found round the northern coasts of Europe, including Iceland, and where it occurs inland in northern and eastern Europe, it lives by large lakes, swamps or rivers. It is slightly larger than the Golden Eagle, with a length of up to 920 mm. and a wingspan of 2,340 mm. and it may be distinguished by its white tail in the adult. Although so large, much of its food is carrion, both fish and other animals, which it may find by water and it only takes large living prey very occasionally. When hunting fish, it uses its talons to grasp those which it can see near the surface and though it may sometimes submerge completely, does not usually do so. In spite of its general harmlessness, it has been mercilessly persecuted over much of Europe and is now extinct in many areas where it was formerly common. In 1968, an attempt was made to reintroduce these birds to Fair Isle, a group of islands to the north of Scotland, which was one of their last British strongholds. Four young birds, taken from

threatened eyries in Scandinavia, were reared and finally released, but their subsequent fate is unknown, apart from one which was found dead soon after. Through most of the Continent it is now a protected species, but in many areas this is difficult to enforce since it is a bird of remote places, far from enlightened surveillance. White Tailed Eagles nest on rocks on the ground or in tall trees, adding to the nest each year, until it becomes a vast, unwieldy structure. The 2—3 eggs are incubated chiefly by the female, but the male remains in attendance and feeds her and guards the nest. The young are fledged at the age of about ten weeks, but are not sexually mature until they are five years old.

BLACK KITE —
Milvus migrans

The Black Kite is a summer visitor to much of Europe, but is absent from the channel coast and has been recorded only as an extremely rare vagrant to Britain. Apart from this, it is widespread over the warmer parts of the Old World, where in many places it is a scavenger in villages and the countryside. However, it is most often found near lakes, rivers or the sea. Here it may hunt like an Osprey, plunging in, talons first, to grab a fish which is near to the surface of the water. On the shore or in estuaries there may also be pickings in the form of carrion washed up on the beach. In many respects the Black Kite resembles the Red Kite but the tail is less deeply forked and the colour is much darker, though it is not black. Most of the European Black Kites overwinter in Africa and very large flocks of birds may gather for their migration flights. It is also sociable in the breeding season. If possible the nests are made in tall trees, but other sites may be used and old nests of other predacious

Haliaeetus albicilla

birds may be taken over. Before mating, both partners perform spectacular display flights, soaring and diving over the breeding area. Between two and three eggs are laid and incubated for about 30 days, mostly by the female. After the young have hatched, the male provides food for them for the first few days, but soon both parents are busy keeping pace with the appetites of their family, and bringing in small vertebrates and carrion.

CORMORANT —
Phalacrocorax carbo

The Cormorant is found round the coasts of Europe and many parts of the tropical Old World, and when it lives inland it is usually found near to large lakes or rivers. It may be recognised by its large size (up to 950 mm. long and with a wingspan of 1,525 mm.) its dark colour, broken by a white chin patch and in the breeding season by a white patch on the thigh, which is visible when the bird is in flight. In Europe a subspecies occurs in which almost the whole head and neck are white, but this is not found in Britain. Renowned for its huge appetite, the Cormorant feeds on fish, which it hunts under water, propelling itself with its large feet and holding the wings close to the body. Fishermen dislike and destroy the Cormorant wherever possible, but it survives partly due to its wariness and partly to the fact that it produces three or four young each season in its untidy nest made of bits of stick and seaweed on cliff tops, ledges or even in trees. Outside the breeding season the birds may be seen on inland waters such as reservoirs. At sea, Cormorants may be seen flying low, just topping the water, or, after fishing, sitting with their wings outstretched on buoys or posts by the water. The reason for this is said to be that they are drying their wings, but why this should be more necessary than with other sea birds which do not have the habit, is unknown. While sitting, they may often be seen to shake their heads, in a quick, irritable manner. They are at this time removing from their nostrils drops of brine, for the Cormorants swallow small fish under water and must actually swallow some sea water. This is desalinated by a gland in the head and droplets of highly concentrated brine are excreted in this way.

SHAG —
Phalacrocorax aristotelis

Closely related to the Cormorant is the Shag, distinguished by its smaller size, (length 635 mm. and wingspan 1,220 mm.) and lack of white on the body. During the breeding season, the Shag has a distinct crest. It is found on cliffed coastlines from Iceland and North Cape down to the Mediterranean. It is not normally seen inland or in estuaries, for it is a more sedentary bird than the Cormorant and does not undertake even minor migrations. Although persecuted to some extent by fishermen, the Shag is increasing rapidly round the rocky coastal areas of Britain. Its nest is always on a remote cliff ledge, and is made, like that of the Cormorant, of sticks and seaweeds. The 3—4 eggs are incubated by both parents for about 31 days. The chicks, which are at first naked and helpless, are fed by both parents, which bring back large amounts of fish in their gullets. The young push their heads into the adult birds' throats to get at the food, which consists mainly of small species of fish, such as Gobies, Blennies and Sand Eels, which are of no commercial importance.

WHITE PELICAN —
Pelecanus onocrotalus

Two species of Pelican are to be found in Eastern Europe. Both are large birds, with a length of 1,650 mm. and a wingspan of 2,500 mm. The species illustrated is the White Pelican, which is rare and declining in Europe, breeding in Bulgaria and Romania and wintering in Greece, but which has its headquarters in northern India and parts of Africa, where it is abundant. It can be distinguished from the Dalmatian Pelican, at least in flight, by the large amount of black on the underside of the wing, whereas the Dalmatian species has very little. In the breeding season the White Pelican has a pinkish tinge to the plumage, and a subspecies, which is found chiefly in Eastern Asia, carries a good deal more pink and is known as the Rosy Pelican. It breeds socially in large colonies in swamps and marshes, using reeds and grasses to make the nest. Incubation of the 1—3 eggs is mainly by the female; at first the chicks are naked, but soon grow a covering of brown down. This is replaced as they fledge by brown feathers, for the young Pelican does not get its white plumage until it is two years old. Pelicans usually fly in lines, with gliding periods between strong wing strokes and they may sometimes soar to great heights. When swimming they sit high and buoyantly on the water. They do not dive, but often fish socially, driving small fishes into the shallows, from where they may be scooped up with ease.

GANNET —
Sula bassana

Gannets may be seen round much of the coasts of Northern Europe in the summer months. These magnificent birds glide on long, narrow wings, which may span 1,730 mm. Ungainly on land, they breed on a few sea cliffs and remote islands in closely packed gannetries. There are 13 of these round the coasts of Britain and two-thirds of the world population of gannets are hatched there, although some are born on rocky cliffs on the American side of the Atlantic. The sexes are similar and courtship elaborate, leading to the laying of the single egg in a nest of seaweed and seashore debris. Incubation is by both parents, who first carefully place their large feet over the egg and then lower themselves on to it. Incubation is lengthy, lasting about 44 days, and the nestling does not fly for another three months after hatching. The first plumage of the young birds is dark, speckled with white; this is lost at the age of four years, when the white adult plumage with the golden head feathers is grown. After the breeding season the adults disperse over the Continental Shelf area, while the young migrate to tropical waters. Gannets feed on fish and to see them hunting is one of the most breathtaking sights of the bird world. They cruise over the water with slow, powerful wingbeats at a height of anything between 10 and 35 metres, and on sighting a suitable fish, half close the wings and plummet down vertically into the sea. The top of the head is protected against the force with which they enter it by specially thickened bones. Strangely, the fish is not speared on the Gannet's sharp beak. The dive gives the bird momentum to swim up under the fish and snap it up as it approaches the surface.

Phalacrocorax carbo

Milvus migrans

Pelecanus onocrotalus

SPOONBILL —
Platalea leucorodia

This large, white relative of the Herons and Storks, which measures 860 mm. in length and has a wingspan 1,370 mm., is to be found in southern Spain and the Balkans, breeding in scattered colonies, often with other species of Heron-like bird. In America the Roseate Spoonbill, which when adult develops a beautiful pink and dark red plumage, nests in the extreme south of the United States and southwards into tropical America. Spoonbills breed by the sea shore, in estuarine areas or in extensive inland swamps. The sexes are similar in appearance and at the start of the nesting season there is a complex courtship ritual, which involves dancing, raising the crest and bill clapping. The nest is usually a platform of sticks in a shrub or small tree. Both parents help in building the nest and in incubating the eggs. This lasts for about 21 days and subsequently they tend the nestlings. These leave the nest at about four weeks, before they can fly, but are surprisingly agile at climbing about the branches of the nesting trees. After breeding the European birds migrate southwards, taking a western coastal route, while the birds from Eastern Europe go directly southwards to North Africa. In flight they may be distinguished from herons by the way that the neck is held extended, rather than tucked back into the shoulders. Spoonbills' food consists of a wide range of small water plants, fishes, tadpoles, amphibian spawn, aquatic worms and insects and their larvae. These are caught by the bird as it wades through shallow water, swinging its beak from side to side, filtering out its food as it goes.

Sula bassana

LITTLE EGRET —
Egretta garzetta

In parts of Spain, the south of France and the Balkans, the Little Egret may be met with in swamps, river deltas and marshes, where it nests colonially in bushes overhanging the water. It is reported very occasionally from estuaries in south and east Britain, but these are vagrant birds, which do not breed. It is, however, one of the most widespread of all the species to occur in Europe, for it is found across the Old World from Spain to Australia. The only bird with which it might be confused is the Great White Heron (see page 101), which is larger than the Little Egret's 560 mm. length and 980 mm. wingspan. Also it is unlikely to be met with outside the extreme south east of Europe. The Little Egret has black legs and distinctive yellow feet and in flight holds its neck in a tighter S-bend than does its larger relative. The elongated shoulder plumes, for which these birds were once hunted, are present only in the breeding season. The nest is made mostly by the female, from materials supplied by the male. Both parents incubate the eggs and feed the young on a wide variety of small fishes, amphibians, molluscs, insects and grubs. Outside the time when young are being reared, the Little Egret migrates southwards and is most likely to be seen in areas of brackish water in lagoons and estuaries.

CATTLE EGRET —
Ardeola ibis

Slightly smaller than the Little Egret, but unlikely to be confused with it because of its stockier shape and the golden buff of its head and back that it develops in the breeding season, is the Cattle Egret. It often nests with the Little Egret, but its feeding behaviour is quite different, for it tends to go to dry land to follow cattle and pick up the insects which they disturb. Although resident in Europe only in southern Spain and Portugal, it is also found through much of Africa, eastwards to Madagascar and in Central and Southern Asia. In these areas it often uses large game animals in much the same way that it does cattle in Europe and may often be seen with Elephants or Buffaloes. In recent years it has been extending its range, first by crossing the Atlantic into South America, from where it has spread northwards. It has been introduced in Australia and is thriving there. It seems to be a great traveller; parties of birds, or single individuals may be met with well away from their usual haunts, such as those occasionally recorded from Britain.

FLAMINGO —
Phoenicopterus ruber

This species of Flamingo is abundant in Africa and parts of Western Asia, but in Europe breeds only in the Carmargue in southern France and possibly in the south of Spain. They may occasionally be seen all over the Continent, but there is little doubt that the majority of these birds are escapes from zoos and safari parks, where they are always a popular exhibit. When seen, a Flamingo is unmistakeable, for its bright colour, very long legs and neck and strange shaped bill add up to a combination of characteristics possessed by no other bird. In the shallow, saline lagoons of the Carmargue the Flamingoes find the conditions that they need for feeding and nesting, although this is not always successful for a number of reasons and in spite of strict protection the colony is dwindling fairly steadily. The nest, which is on a beach, or in quite shallow water is a curious chimney-like pile of sand and mud in which the single egg is laid. This may be destroyed by many things; as a result of the mistral, waves may build up in the shallow lagoons and inundate the egg or young, or they may be attacked by foxes or Herring Gulls, which will steal both the eggs and the nestlings. Perhaps because of this, Flamingoes are very wary and difficult to surprise, for even when the majority are feeding or seem to be asleep, some individuals will be alert and watching

Egretta garzetta

Platalea leucorodia

Phoenicopterus ruber

for danger. They need a run in order to take off, which they do in a noisy pink cloud. They are powerful but elegant in flight, with both legs and necks stretched out to give them a cruciform outline. In spite of their leg length, Flamingoes incubate their eggs normally, and after about 30 days, the chick, which is at first covered with grey down, hatches. Its beak is straight and the colour of the legs differs from that of the parents. It is soon able to leave the nest and the young form huge crêches, attended by only a few adults; at the age of ten days it can swim, if need be. It is fed with semi-liquid food by both parents for about 70 days, by which time it can fly and feed itself. The grey juvenile plumage is moulted for adult colours at the age of about one and a half years, but the young birds are not sexually mature until several years after this. The feeding mechanism of Flamingoes is different from that of any other bird. The long neck is twisted so that the beak is held upside down in the water, which is drawn in by pumping actions of the fleshy tongue. Small organisms, such as molluscs, crustaceans and algae are filtered by horny plates which hang down from the upper mandible. In the species of Flamingo, shown in the illustration, these plates or lamellae are discontinuous, so a wider variety of larger food may be taken.

SHELDUCK —
Tadorna tadorna

The 600 mm. long Shelduck may be seen on mud flats and estuaries round the coast of Britain and much of Europe from north Norway to Brittany and also in a few places in the Mediterranean and Black Sea where they breed. In the wintertime thay may occur from Britain southwards, for they migrate to the warmer parts of the Continent from the cold areas of their range. They are among the most easily recognisable of ducks, for both males and females have the bold colour pattern shown in the illustration, but the drake can be distinguished by the large red knob at the base of his bill. In many respects the Shelduck are goose-like in appearance, but no European Geese are as brightly coloured. A name sometimes given to the species is the Burrow Duck, for the nest is made in the shelter of a rabbit burrow or some other hole.

When hatched the young often congregate in large numbers with a few adult 'nursemaids'. They feed on a wide variety of small invertebrates and a little plant food is also taken.

RUDDY SHELDUCK —
Casarca ferruginea

This handsome, mainly chestnut coloured duck, breeds in a small area of southern Spain, in part of the Balkans and the extreme east of Europe and West Asia. It is a popular bird in waterfowl collections, and records for the northern part of the Continent are almost certainly of escaped specimens. Like the previous species, which it resembles in its goose-like characteristics, it nests in burrows and feeds on small invertebrates.

EIDER DUCK —
Somateria mollissima

The specimen illustrated is the Eider drake. His mate is cryptically coloured in browns and greys, which disguise her as she incubates her eggs. Eiders are birds of northern waters, nesting near the coasts of Iceland, Scandinavia, northern Britain, Denmark and part of Brittany. When not nesting, they are at sea, usually beyond the range of breaking waves. Here they dive for their food which includes molluscs, crustaceans and sea urchins. At the start of the breeding season small flocks of Eider may be seen not far from the land and the crooning voice of the drake may be heard. During the courtship ceremonies, the female points with her beak at a male on which her mate then makes a (usually half hearted) ritual attack. The female makes the nest which she lines with down feathers preened from her body, a sacrifice which is not as extreme as it is often pictured, for she is moulting at the time. The down protects the eggs and helps to insulate them. In some areas, particularly in Iceland, the Eiders are more or less farmed; they are provided with suitable nest sites and the down is then removed, for it is valuable as the filling for quilts and pillows. Two crops can be taken, but the third time the female makes her nest she is allowed to keep it, for by then, her supply of down is nearly exhausted and she must have enough to protect the eggs. She incubates them alone for about 24 days, during which time she does not feed and leaves the nest only on two or three occasions for a short time, when she needs to drink. As soon as the ducklings are dry after hatching, they are led down to the sea by their mothers. They may have to hop down quite steep cliffs to reach it, but usually do so successfully. They then join the flocks which include the males, although it is their mothers which continue to care for them, for it is some time before they are fully feathered and can fly.

DUNLIN —
Calidris alpina

The Dunlin is a small wader, about 175 mm. long. It is the most common wader of European shores, sometimes occurring in flocks of tens of thousands of birds. One of the most exciting sights of winter bird watching is to see one of these great groups taking off and flying round in perfect formation, looking from the distance like a low, fast-moving cloud. Dunlin breed on high moors in Iceland, Scandinavia, parts of Northern Europe and upland Britain. They produce four fluffy young which can soon leave the nest, but continue to be cared for by both parents until they can fly at the age of three weeks. Dunlin have fairly short legs and short beaks, so their diet is restricted to small invertebrates which they can catch in shallow water, or even obtain on dry land.

RUFF —
Philomachus pugnax

While it is in general true to say that no two animals are identical, this is perhaps more apparent in the male Ruff than in any other creature. During the breeding season these birds develop large frills of feather round the neck on the top of the head, which are individually different in colour, pattern and size. This elaborate neckwear is part of the sexual display system, for the male Ruffs congregate at a dancing ground or 'lek', where they display to the females, indulging in mock battles during which the neck plumage is erected. Both males and females are promiscuous, and the female incubates her four eggs and

rears the chicks alone. The breeding grounds of the Ruff are from the northern Low Countries northwards and east into Siberia, mainly on high moorlands and tundra. At one time they bred in Britain, but since the middle of the last century they have been seen only as passage migrants. Outside the breeding season the males loose their picturesque display feathers, but groups of Ruff, which tend to be seen more on inland marshes and meadows than on the sea shore, may be recognised, among other features, by the difference in size between the sexes, the males measuring about 280 mm. and the females 230 mm. long. This is unusual, because in most birds where there is a size difference, it is the female which is the larger of the pair. The small, tight little flocks will probably be probing the ground for the worms, grubs and small molluscs which are their food.

Samateria mollissima

Philomachus pugnax

Pluvialis apricaria

Tadorna tadorna

GOLDEN PLOVER —
Pluvialis apricaria

As with many wading birds, the 250 mm. long Golden Plover's summer plumage differs considerably from that worn outside the breeding season, for the dark stripe, running from the beak and eyes to the underparts of the bird disappears during the wintertime. At all times, however, the golden freckles on the back should help to distinguish it from its near relative, the Grey Plover, with which it may often be seen. Another feature which may help to identify the bird in the field is the 'armpit' which is white, not grey or black as in some other plovers. This may often be seen because of the species' habit of stretching up a wing as if in salute, when at rest. Golden Plovers nest in moorlands, where the female rears four chicks, and the male stands guard. In the wintertime they may be met in a variety of habitats, including estuaries and mud flats, where very large flocks sometimes build up. At one time they probably bred further south than they do now, but the improvement of moorlands for agriculture and hunting in the more populated areas has reduced their numbers and they now nest from central Britain northwards, through Scandinavia and into Asia.

OYSTERCATCHER —
Haematopus ostralegus

The piping call of the Oystercatcher, alarmed as it feeds on the beach, is one of the most characteristic sounds of the British coast. Its pied appearance in flight and at close quarters its long, coral red bill and legs make it quite unmistakeable. Its breeding grounds include much of the coasts of Britain and Northern Europe, including Iceland and, in the east of the Continent the borders of the Black Sea and the wet, lowland plains of Western Russia. In the wintertime many migrate to the west Atlantic and Mediterranean coasts. Because they nest at the top of the beach, where an extra high tide may occasionally disturb the eggs, Oystercatchers are among the species which can return their eggs to the nest scrape. This is a rare achievement, for in many cases, once they have been moved, even if only a little way, they are no longer of interest to the bird. They have been the subject of intensive investigation on how birds recognise their eggs, a topic which throws sidelights on many aspects of vertebrate behaviour. Oystercatchers never catch Oysters. They may be seen probing into sand and mudflats and taking worms, crustaceans, and small cockles (see page 274) or venus shells which they lever open with their long beaks.

CURLEW —
Numenius arquata

The musical but melancholy whistle of the Curlew is one of the most frequently heard sounds on moorlands and mudflats in Britain, where the bird is resident. In Southern Europe, it is absent, other than as a winter visitor, for it breeds from the Low Countries eastwards into Siberia. It is the largest of all the European waders, measuring 555 mm. in length and may be easily recognised by its long downcurved bill, with which it probes for worms, grubs and small molluscs. The nest is hidden among the stems of heather and heathland plants and in it the female lays four eggs camouflaged with blotches and stripes. Both parents incubate the eggs and protect the young, which can run about within a short time of hatching. At first their beaks are straight and relatively shorter than those of their parents, but as they grow they change shape. Outside the breeding season Curlew usually congregate in small flocks, sometimes in company with their relative the Whimbrel.

COMMON GULL —
Larus canus

In spite of its name, the Common Gull is not the most abundant species in Britain. It breeds round the coasts of Europe northwards from Holland, including Scotland and Northern Ireland and inland from the Gulf of Finland to Siberia. It also occurs in western North America. It is a medium sized bird, about 400 mm. long, with greenish yellow legs and a yellow bill. It usually breeds in small colonies, on rough or broken ground. Three is the normal number of young: they are tended by both parents and are able to fly at the age of 4—5 weeks, although at this stage they have the mottled brown plumage of the juvenile, rather than the grey and white livery of the adult birds. On their winter migration, Common Gulls fly to Southern Europe and at this time may be seen over most of Britain. They are scavengers, to be found anywhere where food is available, including the sea shore, harbours, and sometimes rubbish dumps. They do not hesitate to steal from the bolder, but smaller Black Headed Gull if the chance arises.

HERRING GULL —
Larus argentatus

This large gull, which measures about 560 mm. in length is one of the most successful birds in Britain today, for its numbers are increasing faster than those of any other species. Its nesting distribution includes European coastal areas where it used to breed at the tops of cliffs and on beaches, but recently it has started to breed on buildings and in marshy places. They are found widely in North America also. Herring Gulls will eat almost anything, from refuse to abandoned eggs and chicks of their own species. A trick which it shares with the Common Gull is that of dropping molluscs such as cockles (see page 274) which it has dug out of the sand on to hard surfaces such as rocks, or even roads to break the shell. These large, bold birds are easy to watch. Perhaps because of this they have been made the subject of intensive behavioural investigations, which are of value in our understanding of all vertebrates.

ARCTIC TERN —
Sterna paradisaea

While some terns, such as the Sooty Tern, often nest inland in swampy places, the Arctic Tern is far more maritime and never nests very far from the sea. This elegant little bird, which measures only about 380 mm. in length, including the streamer feathers of its long, forked tail, is the greatest of all bird travellers It migrates every year from the far north, where it breeds, to the Antarctic Circle, where it spends the Winter — a round trip of about 33,200 kilometres. Its close relative the Common Tern is similar in appearance, but may be distinguished by the black tip to the beak and the slightly longer legs, set more forward on the

Numenius arquatus

Larus canus

body; this bird is often seen further inland. Terns feed on small fishes, such as Sand Eels and when hunting, fly a definite beat of a few metres over shallow water, sometimes hovering for a few seconds before plopping in and usually emerging with the little fish in the beak. Terns nest in densely packed colonies at the top of sandy or shingle beaches. The nest is a shallow scrape

Larus argentatus

Uria aalge

and the eggs camouflaged to look like pebbles. Incubation and chick care is by both parents until the young fly at the age of 3—4 weeks.

GUILLEMOT —
Uria aalge

The Guillemot is a member of the Auk family, a group of marine birds which are, in general, short-winged and spend a great deal of their time on or under the water, where they hunt small fishes. They swim with their wings and use their feet only for steering, unlike Cormorants (see page 282). In early Spring Guillemots and their relatives the Razorbills come into inshore waters from where they survey their breeding cliffs. Soon after, they occupy the ledges on which the eggs are to be laid. This is done without the benefit of nesting material, on ledges which may be no more than 100 mm. wide. They are strongly pointed at one end; this has the effect of making them swivel on their own axis if a bird accidentally knocks an egg as it lands on or leaves a breeding ledge. Guillemots breed round the rocky coasts of Britain, Northern Europe and Asia and are also found on the coast of North America. When the young are hatched after 30 days incubation, the parents feed it for another 2—3 weeks. Then, although it is far from fully fledged, its waterproof feathers have grown and it flutters from the cliff down to the sea, where it joins the old birds. Here the Guillemots remain until the next year, only coming ashore when wind blown by exceptional storms.

MAMMALS

COMMON SEAL —
Phoca vitulina

All the seals found in European waters have the hindlimbs turned back, so that when on dry land they can take none of the animal's weight. As a result they have to hitch themselves forwards with a caterpillar action of their bodies and forelimbs which makes them look fat, heavy and ungainly. The picture is reversed when they are in the water, for here their streamlined shape enables them to swim with ease and grace, using the hindlimbs in a side to side motion and the front ones merely to steer or balance, or to paddle with at very low speeds. They are insulated against the cold with a heavy layer of fat and are able to hold their breath for long periods while they hunt. The Common Seal is sometimes called the Harbour Seal, an appropriate name, for wherever it occurs it is an inshore species. It is usually associated with lowland shores, sandbanks and mudflats rather than the cliffs and caves favoured by the Grey Seal which is the other species usually found round Britain. The Common Seal is found in all the seas of the northern world, as far south as California in the Pacific and as France in the Atlantic. It is a small species, the largest of the males being less than 2,000 mm. long and the females are much smaller. It is very variable in colour. It normally lives in small herds, usually not travelling very far from a particular living place, although young Seals may migrate long distances and are sometimes recorded far up large rivers. Each female Seal produces only one very well developed young each year. The baby is able to swim within hours of its birth and remains with its mother for several weeks, probably fairly close to the place where it is born. During this time it is suckling and this is also the time when Seal hunts take place as for example in the Wash on the east coast of England. Many of the young may be killed. In more northerly areas they may be hunted for food at any time, particularly during the Winter. These Seals feed on flat and other bottom living species of fishes. Apart from the Killer Whale, man is their only enemy.

BOTTLE-NOSED DOLPHIN —
Tursiops truncatus

Dolphins are small relatives of the great Whales and like them are mammals, air breathing and warm blooded, producing and suckling living young. The Bottle-nosed Dolphin is found widely through the Atlantic Ocean from Bear Island to the Mediterranean and also in American waters. Its size varies from under 3,000 mm. to over 4,000 mm. but it is not the smallest of the Cetaceans, for its cousin the Common Porpoise, which is more abundant in British waters is less than 2,000 mm. long. Bottle-nosed Dolphins are probably the more familiar species, however, for they are the kind most often kept captive in aquaria. Like most Whales they are highly sociable animals and live in schools which may, on occasion, contain several thousands of individuals. They are certainly highly intelligent, and tests on captive specimens have led some scientists to suggest that after man they are the cleverest of all animals. In spite of the fact that they have been hunted almost to extinction in the Mediterranean and are often persecuted elsewhere because they feed on fish, they seem to bear man no malice and there are many stories, both from antiquity and modern times of Dolphins raising shipwrecked mariners to the surface of the water and thus saving their lives, which is something that they will do to members of their own kind in distress. On many occasions Dolphins have taken to accompanying boats or swimmers near the shore. They often swim with larger vessels at sea, sporting in the bow wave and keeping up speeds of up to 25 kilometres per hour for long periods. Although they have no external ears, they make a variety of squeaking noises which are apparently part of an echo-location system. This enables them to avoid obstacles and find their prey even in dark or murky waters. As well as this they communicate with each other and some people hope that one day man may learn to understand their language.

Phoca vitulina

Tursiops truncatus

IN
THE
FAR
NORTH

INTRODUCTION

The far north of Europe may be considered to include most of Scandinavia and parts of Northern Russia and also a number of islands, such as Iceland, Spitzbergen and Jan Mayen. Parts of these include upland areas where there may be glaciers and icefields, but in the lowlands, tundra covers the ground. Beyond this to the south, there are forests, mainly of birch and pine, which house and protect many sorts of animals.

Living conditions in the far north are extremely harsh. The low temperatures and strong winds of the winter months make survival difficult for all but the toughest and best adapted of animals. Many which are resident during the summertime migrate to milder areas when the ground freezes. Others which remain manage to find sheltered spots where food may still be available, in some instances under the blanket of snow, while yet others hibernate through the worst of the weather. These animals which stay active through the Winter have to contend with darkness as well as cold, for north of the Arctic Circle there is a short period when the sun does not rise above the horizon and even when it does, the day length is very short indeed. This is compensated for in the summertime, when daylight may be present throughout the twenty-four hours, but even at the height of the warm season, the temperatures are low by comparison with the rest of the Continent.

Much of the far north is covered by a type of vegetation called tundra. This consists mainly of dwarf woody plants, some species of which may be only a few centimetres high, although their relatives of temperate regions may be large trees. A few tundra species are found in upland areas of northern Britain, as relicts of the Ice Age. Willows and birches in these places are shorter than the meadow grasses of more temperate areas. Mosses and lichens also abound in the tundra, and may in some cases be important food sources for large animals. Some of the lichens in particular are very nutritious and a few are also pleasant to eat. Children of the high Arctic may search for these species in the same way as city children keep their eyes open for shops selling ice cream.

The ground beneath the tundra plants is permanently frozen, which is one of the limiting factors to their size. In the warmest months the surface may thaw to a depth of one metre, but it is often much less than this. Snow melt water cannot soak through the ice, so the tundra is characterised by many small, shallow lakes which form in hollows. These are ideal breeding grounds for many types of aquatic invertebrate, especially those related to the midges and mosquitoes, and a visit to the Arctic can be a nightmare of insect bites. Yet it is these same insects which are largely responsible for the huge summer influx of birds to the far north. Many waders and other small birds find that the long days give extra working hours for stuffing the gaping mouths of their nestlings with the abundant, easily available food. These birds, however, are only long-distance commuters and they return to more congenial climates once they have reared their brood.

In spite of the complexity of inter-relationships between climate, soil and plants, the Arctic still offers a poor environment which is relatively much simpler than that of warmer areas. Perhaps because of this, it is easier for a single species to get out of balance with its habitat controls. In many species there are periodic huge fluctuations of population, the best known being that of the Lemmings. However, other species including some small mammals and birds, such as Nutcrackers and Waxwings, show similar if less dramatic variations in numbers, and like the Lemmings they may then spread far beyond their normal environment. At the time of such irruptions, predators, such as Owls and Arctic Foxes increase in numbers also.

In the far north the sea is more evident as a source of life than it is further south. The mineral richness of the water gives rise to huge spring 'flowerings' of minute sea plants which are the food for many small invertebrates. These in turn nourish fishes, birds and mammals. The huge flocks of Great Auks which occupied these seas were virtually extinct early in the last century and the mighty Whales which survived as relatively common animals into the beginning of the present century were both dependent on this wealth. Today it is the sea which nourishes countless cliff-nesting marine birds. Auks, Fulmars and Gannets are found and Tubenoses, related to the Albatrosses of the southern oceans, are abundant in some places, while Gulls scream and scavenge throughout the area.

An interesting observation made on the fauna of the far north is its present tendency to move southwards. In recent years this has been most clearly demonstrated by the several species of northern birds, including the Redwing, Great Northern Diver and Snowy Owl which have all nested in Britain for the first time. Changes have also been noted in the microfauna of the oceans, but whether these phenomena herald the onset of a new ice age, or are due to some other cause is not yet apparent. However, the trend is being watched with interest by zoologists and naturalists.

INSECTS

Aedes sp.

Any visitor to the northern parts of Europe, Asia or America, will be attacked, during the summer months, by countless biting flies. These may be closely related to the mosquitoes or midges found further south, but probably belong to different species. Culicine mosquitoes form one of the commonest groups, recognised in the adult by the way in which they sit with the body parallel to the surface below them. An important genus, of which 14 species occur in Britain, is *Aedes*.

These insects lay their eggs in compact rafts on water, where the larvae develop. The food of the adults varies between the sexes; the male, which can be recognised by his feathery antennae, sucks plant juices; the female feeds on blood. This is necessary to her before she can lay her eggs, and she will bite any mammal, including man. Even Elks are plagued by them and try to escape by spending long periods almost completely submerged in pools or rivers.

BIRDS

SNOWY OWL —
Nyctea scandiaca

The Snowy Owl is a bird of the high Arctic and tundra, where it feeds on Lemmings, Arctic Hares and even Eider Ducks. Its large size — the female is up to 650 mm. long — and its almost completely white colour should make identification easy, but the large round head, staring yellow eyes and heavy 'moustache' are further characteristics unlike those of any other bird. It is more active during the daytime than are other species of Owl and often sits on a haystack or rock to survey the surrounding low vegetation for suitable prey. At other times it flies and glides slowly over the ground, scanning it carefully for food animals. The breeding season starts in late Spring but the first egg may not be laid until June, in a nest which is usually in a burrow or other sheltered spot on the ground. In years when the Lemming populations are high,

the Snowy Owls thrive and may rear large families. At other times only one or two young may be fledged. As a result, population fluctuations following those of the prey animals occur, and Snowy Owls especially young ones, recognisable by their darker colour, may sometimes be seen well to the south of their usual range, which includes parts of Iceland and northern Scandinavia. In 1967 Snowy Owls nested on the island of Foula, in the Shetland Islands. This was an important ornithological event, for they had never in historic times been recorded as nesting anywhere in Britain. These birds, which were carefully observed, fed mainly on rabbits. They have bred in the same area in subsequent years, but have not yet spread to any other southern locality.

GREAT SKUA OR BONXIE —
Stercorarius skua

At first sight the Great Skua looks like a stoutly-built brown gull, but in flight its obvious white wing patch will enable it to be identified easily. Although related to gulls, Skuas lead a very different kind of life, reflected in the name 'Jaeger' by which the group is known in America. This is derived from the German word for a hunter, although it might perhaps be more

Nyctea scandiaca

Fratercula arctica

Stercorarius skua

They will feed on the carcasses of dead Seals or Whales cast up on the shore and have discovered, along with the Fulmar, that fishing boats in northern waters mean a rich harvest of offal for them. Perhaps because of this new food supply, their numbers, like those of the Fulmar, have been increasing during this century. They now breed on the Scottish mainland as well as in Iceland, the Faeroes, Shetland and Orkney. During the Winter they take to the ocean and may be seen as far south as the tropics. However, this is not the furthest south that the species extends, for in the Antarctic the Great Skua scavenges Penguin colonies and kills any weakly chicks, as well as following its usual piratical way of life.

PUFFIN —
Fratercula arctica

The distinctive appearance of the 300 mm. long Puffin is familiar to many people who do not know the actual bird, for it is the trade mark of various commercial goods. To see it, it is necessary to visit the coasts of North West Europe where these birds breed during the summer months. Making their living from the sea, they seem, like their relatives the Guillemots (see page 290), reluctant to come ashore. When they do so, they choose grassy slopes at the top of steep cliffs to make their nests, for although they can escape enemies at sea, they are more helpless on land, so the eggs are laid and the young reared in burrows. These may be dug by the parent birds, which use their large beaks to loosen the soil and stones which they then kick out with their webbed feet, or they may take over and refurbish disused rabbit burrows. During the breeding season the Puffin's beak thickens and develops extra colour, for it is used in courtship display and in aggression to rivals for mates or living space, as well as the usual purposes of feeding and preening. After the young are reared, the bill sheds its gaudy outer layers and the white facial feathers are moulted for others of a dingy grey colour. Puffins may line their nest burrows with vegetation or feathers, but each pair produces only one egg. This is incubated for about forty days, mainly by the female, although the male later helps her to feed the chick. It is at this time that Puffins may be seen with a beak-

accurate to call them pirates. The Great Skua, which is about 580 mm. long harries other sea birds until they drop the fish that they have caught. In pursuit of such a meal the Skua is a skilful flier; no matter how its prey twists and turns, the big bird can keep up. They will even attack Gannets (see page 283) which are far larger birds, and have been seen grabbing a wingtip or tail until the Gannet regurgitates its fish, which is caught before it hits the water. They always seem to know which birds have recently caught fish and neglect those which have been unsuccessful. The sight of a Skua will throw a flock of Terns into a panic, for their young may sometimes be killed by this marauder. In defence of their own young, which are produced in nests on the ground, that may be solitary or in scattered colonies, the Skuas are even bolder than most birds. They will not hesitate to buzz a human intruder, flying at him fast and low, swerving at the last instant, or perhaps, with a persistant intruder, the bird will deliver a sharp blow on the head or shoulder as it speeds past. Great Skuas make a living by scavenging as well as by piracy.

ful of tiny fishes, neatly arranged crossways, held against the upper bill by the tongue, so that more can be caught if need be. Gulls often lurk about Puffin colonies, trying to scare the returning birds into dropping their catch. Since they will not face the large bill at close quarters, once a Puffin reaches its burrow it is safe. The young are deserted by their parents after about six weeks of care. During this time they have been fed to repletion until they are large and fat and weigh more than the adult birds. For about a week they remain unfed, using the reserves of food for final feather growth until eventually they leave to join the flocks of old birds at sea. Puffins are becoming rarer over much of their range. The reasons for this may be complex, but one major cause is oil pollution of the sea, for these birds, like all members of the Auk family dive and swim underwater. They may come up through a patch of oil, which they cannot detect from below, but which, sticky and poisonous, spells sure death for them once it has contaminated their plumage.

MAMMALS

BLUE HARE —
Lepus timidus

Many subspecies of the Blue Hare exist, found in the northern and some mountainous parts of the Old World. In Britain, the Blue Hare occurs in the highlands of Scotland and in Ireland, where the Brown Hare is absent, apart from recent introductions. The Blue Hare is somewhat smaller animal than the Brown Hare, with very much shorter ears, but an easier point of distinction is its totally white tail, and the tendency for the coat colour to turn white during the wintertime. It seems to be more active in the daytime than the Brown Hare and also to swim readily if need be. The lying-up place, or form, is usually in a sheltered spot between rocks, or may even be a tunnel which has been excavated for a short distance. Blue Hares are somewhat more social than Brown Hares, but the density of the population in Scotland varies greatly from one year to the next, though not so dramatically as its close American relative, the Snowshoe Rabbit. The food of the Blue Hare includes many sorts of tundra and mountain plants. In wintertime it may move from high ground to more sheltered places, though even here the vegetation may be covered with snow, which the Hare scrapes away with its forepaws. Its main enemies are man, eagles, wild cats and foxes.

NORWAY LEMMING —
Lemmus lemmus

Small, mouse-like rodents have colonised habitats throughout Europe, including those of extreme dryness and heat in the south and extreme cold and humidity in the north. The most famous of these northern species is the Norway Lemming, an animal with a body length of up to 150 mm. and as with all Arctic inhabitants a very short tail, measuring at the most 20 mm. It is more brightly coloured than most of the Voles, with an unmistakeable pattern of black on the orange fur, although the exact shade is very variable. It normally lives in the tundra or high mountain zones in Norway, entering the birch and sometimes the conifer zones on the hillsides. It makes extensive tunnel systems under moss and stones and, in the Winter, under snow. It is active mainly at night and does not hibernate during the coldest weather, nor does it make food stores against times of hardship. Its normal gait is a scuttling run and though it is a poor climber it can swim well. It is quarrelsome and noisy with a wide vocabulary of squeaks, squeals and grunts.

Lepus timidus

Lemmus lemmus

As with many other small, northern gnawing animals, the population fluctuates on a fairly regular basis, with periods of abundance alternating every 3—4 years with periods of scarcity. Plague numbers build up to a point that cannot be supported by the environment and many animals migrate away from the centres of population. They swarm downhill, often swimming rivers and broad lakes and some may reach a temporary haven, although they probably do not survive long out of their true habitat. Others, however, go on until they reach the sea which they may try to swim, an action which results in their death. They have no suicidal urges, as is often stated, but are merely looking for an uncrowded place in which to live. At times of Lemming plagues, they may be found throughout Scandinavia. Foxes, Owls and other predators all benefit from their abundance and bewildered behaviour, but after a 'Lemming year' numbers drop disastrously and for a few seasons these little animals are seen only in their upland homes.

GLUTTON OR WOLVERINE —
Gulo gulo

Up to 820 mm. long and standing not more than 450 mm. at the shoulder, Gluttons were at one time classified with bears, to which they have a superficial resemblance. In fact, they are members of the Weasel family, a group whose members are more varied in their looks and ways of life than any other of the carnivores. It is to be found in the mountain forests and the tundra of Scandinavia, Northern Europe, Asia and North America. It is generally a solitary animal, making its den among boulders or at the base of a fallen tree, where two young are normally born to the female early in the year and are cared for by her alone. It can climb, but rarely does so and normally lives on the ground. Active by day or night, the Glutton lives up to its name, for there is little that it refuses to eat. It may hunt or ambush any small animal of its environment, and will also take any carrion that it comes across. This habit in particular has tended to make it unpopular with man, for it lives in areas of intensive fur trapping and will take animals from traps and snares, and like most carnivores it comes under suspicion of stealing chickens and young domestic animals. Apart from this, Gluttons may eat fish, fruit and berries. They may be hunted for their fur, but as this is shaggy and coarse, it has no great value.

ARCTIC FOX —
Alopex lagopus

The Arctic Fox is a small, dog-like animal of the high Arctic, occurring from the coast of Scandinavia, along the north coast of Europe, Asia and North America, and also on Iceland and Greenland. It will willingly travel on sea ice and so has crossed to and colonised offshore islands, even those as remote from the mainland as Jan Mayen and Spitzbergen. It is a smaller

Thalarctos maritimus

Gulo gulo

creature then the familiar Red Fox, with a body length of up to 770 mm. and a tail length of well under 500 mm. A big specimen will stand 400 mm. at the shoulder. It has a shorter muzzle than the Red Fox and small, rounded ears, which are less vulnerable to frostbite. Two colour forms are known: one is a purple-brown in Summer and pure white in Winter. The other, known as the Blue Fox is smokey grey throughout the year. This form exists in most populations, but tends to be more heavily hunted for its beautiful pelt. The Arctic Fox is more nomadic and more social than the Red Fox. It will go wherever there is a promise of food, moving for the summertime to the base of sea bird nesting cliffs and taking the carrion young which have fallen from their nursery and then moving on to the tundra for the Lemming harvest. If food is plentiful they will store it and Arctic explorers have described how they have seen them dig up meat tins from an expedition refuse tip going off to bury them elsewhere. In the Winter they tend to move towards ice-free coasts, where they live on carrion and molluscs or they may move to the shelter of forests. Groups may form systems of runs and burrows, occupied by several families. The females normally produce two litters of from 5 to 8 cubs a year and it may be this fecundity which enables the Arctic Fox to survive the heavy pressures of hunting it for its fur, although it has been noted that periods of abundance tend to follow the years of lemming plagues.

POLAR BEAR —
Thalarctos maritimus

This is another circumpolar species, which may be considered to be a marine rather than a land animal, for only in Iceland does the Polar Bear venture inland any distance from the coast. It may sometimes be seen swimming powerfully, if rather slowly, many kilometres from land and may travel huge distances on ice floes. There are records, although none is recent, of Polar Bears making a landfall in Scotland. Polar Bears could not be confused with any other animal. Their very large size sets them apart from other creatures, except for Brown Bears, from which they differ in colour and the shape of the head and ears, which are small. The male Polar Bear may have a

head and body length of up to 2,500 mm. and stand up to 1,400 mm. at the shoulder, although females are only about two-thirds of this size at most. They have very large feet, densely padded underneath with fur, which not only acts as insulation, but also helps them not to slip on ice or packed snow. In spite of the size of the adults, Polar Bears are tiny when they are born. The females go into a state of hibernation in a snow cave which they hollow out, and their young, usually only one or two to a litter are born in February. Weighing only 600 grammes, they are naked, blind and helpless, but stay close to their mother, taking little but warmth from her, for although she suckles them, their needs are very small. They are able to emerge in March, but stay with their mother for at least a year, so breeding, as with most large mammals, is slow. They quest for food at almost any time of night or day, but are almost entirely carnivorous, mainly hunting Seals of all kinds. These are usually attacked when they are hauled out on the ice, for once in the water the Polar Bear is no match for the Seal. Bears will also wait by a Seal's breathing hole and catch it as it comes up for breath — a technique also employed by Eskimos when seal hunting. Young Walrus may be attacked, but the adults are usually left severely alone. Polar Bears may sometimes congregate to feast, for although they are usually solitary animals, if food is plentiful, they will not insist on territorial rights, but will ignore each other. Several may, for example, be attracted to the carcass of a stranded Whale and for a short time will share the booty. Stranger, however, are the semi-permanent populations of Polar Bears which scavenge the rubbish tips of certain Alaskan towns. Few things are left untried, and the waste from Anchorage in Alaska, for example, is put to good use by the bears. In some areas Polar Bear watching is a local pastime, on a par with bird watching. The bears seem to know that they have nothing to fear and are incurious and non-aggressive towards the watchers. Not all bear-man relationships are so peaceful and in remote places, where food is short, bears may attack human camps. In the past, Polar Bears were widely hunted for prestige and for their shaggy pelts, until their numbers were reduced to a very low ebb. While Eskimos are still generally permitted to hunt them, in many areas they are, apart from this, given complete protection. However, in a few places, in spite of the protests of conservationists, they may be hunted from helicopters or light aircraft.

WALRUS —
Odobenus rosmarus

Since the North Pole is an ocean area, a number of aquatic mammals may be found on the fringes of the ice and the north shores of the great land masses of Eurasia and America. The Walrus is one of these. With its great tusks, which are hugely enlarged canine teeth, and wrinkled, furless skin, the Walrus could not be mistaken for anything else. A relative of the Seals, it is larger than any of them, the male, which is much bigger and heavier than his mate weighing up to 2,200 kg., although the females may weigh as little as 700 kg. Unlike the true Seals (see page 290), the Walrus can turn its hind flippers under the body, which makes them less helpless on land, although they are still very clumsy. They swim fast, using the hind flippers and sometimes the front ones as well and may dive for up to ten minutes and descend to a depth of 30 metres when hunting for food. This consists entirely of molluscs, crustaceans and sea urchins which are dredged from the sea bed with the tusks; then are manipulated into the mouth by the mobile, whiskery lips. Walrus pups are born singly and are cared for by their mothers for over a year. Some Walrus herds are still quite large, but severe over-hunting has much reduced their numbers. In some areas they are now totally protected apart from small scale exploitation by Eskimos, but a few are still hunted mainly for their heavy, but remarkably flexible hide, which has various industrial uses.

ELK —
Alces alces

The Elk is the largest species of deer, a large male having a head and body length of up to 2,900 mm. and standing 2,100 mm. at the shoulder. Females are somewhat smaller, and do not carry the spreading, dish-shaped antlers borne by their mates. The name of this animal causes some confusion. In North America it is called a Moose and the name 'Elk' refers there to the Red Deer. They are indeed related to Red Deer, but have many unique features, including the extraordinary heavy profile, the very high shoulders and sloping back, which serve to identify the species at any time. Elk lived in Britain in late prehistoric times, but were extinct there before the Roman invasion. Today they are found through Scandinavia, North Eastern Europe, North Asia and the northern parts of North America. They are animals of marshy forest country, where in the summer time they spend much of their time by and in the water. They feed by day or night on waterside plants, or the stems of water lilies, which are a favourite food. In the Winter, they may be forced to make southward migrations, when they feed on many sorts of plants, including the shoots and twigs of coniferous trees, and if the weather is particularly bad, they may be forced to eat the bark as well. Apart from migration time, when loose herds may be formed, the Elk is usually a solitary animal, occupying a relatively small territory. The mating season is in September, when the bulls, which are then in the peak of condition after good summer feeding, challenge and fight. They do not accumulate a harem, like the Red Deer, but mate with a succession of females in the area. The young, usually twins, are born in the next May and June and remain with their mother until the subsequent Spring. The young females may mate for the first time at the age of about 18 months; the males are not sexually mature until a year later. Bears and wolves are the chief animal enemies of the Elk, taking a high toll of the young, but they are not such a severe predator as man, who has reduced its numbers everywhere throughout its former range. Hunting is now strictly controlled on both sides of the Atlantic, and the species seems to be holding its own and even increasing in numbers in some areas. Attempts have been made to domesticate the Elk, mainly as a beast of burden, and more recently, in Russia, for milk and meat.

REINDEER —
Rangifer tarandus

Reindeer used to be present as truly wild animals in Britain, but like so many large, edible creatures, were hunted to extinction there before the end of the twelfth century. Now they exist in the wild only in northern-most Europe, Asia, and in North America where they are known as Caribou. In America they remain as wild animals, hunted in some cases by man; in Europe and Asia most are domesticated, although they lead, to a large extent, a life more comparable with that of wild animals than is usual in domestication. They are large creatures, with a head and body length of up to 2,150 mm. and a shoulder height of up to 1,200 mm. They are distinguished from all other deer by the presence of antlers in the females. These are retained through the Winter, although those of the males are cast in the Autumn after the rut or mating season. A single calf is born in the late Spring and after this the females shed their antlers. The calf is weaned and new antlers are grown by the next rut. In the wild, Reindeer inhabit northern uplands, tundra and open forest, especially where there is a good deal of standing water. Perhaps as part of their adaptation for this wet habitat, their cloven hooves spread to take their weight and click together again as the foot is lifted. A herd of Reindeer, especially if moving over hard ground, make this distinctive hoof noise. In general, the Reindeer is a so-cial creature, the females and young forming herds of

thirty or more; the males are solitary during the sum-mer but join the females during the rut and follow the herds in the course of any winter migration. They feed on a wide range of grasses, sedges, broad leaved plants and in the winter on various lichens, one of which is commonly called reindeer moss. The females are reputed to use the inwardly pointing brow tine of their antlers to clear the snow off this fodder. Rein-deer have been domesticated in northern Europe and Asia for many centuries. It is, however, a different form of domestication from most, with the humans being, in effect, parasitic on the herds of deer. The migrations of the deer dictate the travels of the humans, who follow them, half taming them for beasts of burden or for riding, milking them, or killing them

Alces alces

Rangifer tarandus

305

for meat or hides. The lives of the northern Lapps are entirely dictated by the yearly cycle of the Reindeer. Some selection is imposed on breeding, so colour forms, particularly white, tend to occur. In 1952 a small herd of Reindeer was imported into Britain. After initial difficulties, probably caused by their being kept in parasite-ridden lowlands, the herd is now thriving in the Cairngorm area and visitors to that part of Scotland may see this recently returned species living on the upland slopes.

INDEX OF SCIENTIFIC NAMES

INDEX OF COMMON NAMES